George Bell

A catalogue of books of art, natural history, medicine, history, biography, philosophy and theology

George Bell

A catalogue of books of art, natural history, medicine, history, biography, philosophy and theology

ISBN/EAN: 9783742854872

Manufactured in Europe, USA, Canada, Australia, Japa

Cover: Foto ©berggeist007 / pixelio.de

Manufactured and distributed by brebook publishing software (www.brebook.com)

George Bell

A catalogue of books of art, natural history, medicine, history,

biography, philosophy and theology

A

CATALOGUE OF BOOKS

OF

ART—NATURAL HISTORY—MEDICINE—HISTORY—

BIOGRAPHY—PHILOSOPHY—

THEOLOGY,

ETC.

TOGETHER WITH A LIST OF

BOOKS FOR YOUNG PEOPLE;

ALSO VARIOUS APPROVED EDITIONS OF

STANDARD ENGLISH WORKS.

PUBLISHED BY

GEORGE BELL AND SONS,
YORK STREET, COVENT GARDEN,
LONDON, W.C.
DEIGHTON, BELL AND CO. CAMBRIDGE.
1874.

CONTENTS.

GEORGE BELL AND SONS'

CATALOGUE OF BOOKS.

𝔄rtistic and 𝔒rnamental 𝔚orks.

MEMORIALS OF WEDGWOOD. A Series of Plaques, Medallions, Cameos, Vases, &c. Selected from Various Private Collections, and executed in Permanent Photography by the Autotype Process. With Introduction and Descriptions by Eliza Meteyard, Author of "The Life of Wedgwood." Imperial 4to. Handsomely bound. £3 3s.

WEDGWOOD AND HIS WORKS; a Selection of his choicest Plaques, Medallions, Cameos, Vases, and Ornamental Objects, from Designs by Flaxman and others, reproduced in Permanent Photography, with a Sketch of the Life of Wedgwood, and of the Progress of his Art Manufacture, by Eliza Meteyard, Author of "The Life of Wedgwood." Imperial 4to. With Illustrations, handsomely bound. £3 3s.

THE WORKS OF CORREGGIO AT PARMA; Reproduced in Photography by Stephen Thompson from the Celebrated Engravings by Paolo Toschi. With Biographical and Descriptive Notices by Louis Fagan, Department of Prints and Drawings, British Museum. Imperial folio. Half-bound. £5 5s.

TITIAN PORTRAITS. Seventeen Photographic Reproductions of Rare Engravings after Titian in the British Museum, by Stephen Thompson. With Descriptions by G. W. Reid, Keeper of the Prints and Drawings in the British Museum. Large folio. Half-bound. £5 5s.

THE WORKS OF VELASQUEZ. Seventeen Photo-graphs, by Stephen Thompson, of Scarce and Fine Prints in the British Museum. Selected and described by G. W. Reid, Keeper of the Prints and Drawings. Large folio. Half-bound. £5 5s.

THE CASTELLANI COLLECTION; a Series of Twenty Photographs by Stephen Thompson, selected and de-scribed by C. T. Newton, M.A., Keeper of the Greek and Roman Antiquities, British Museum. Folio. Cloth binding. £3 3s.

THE WORKS OF WILLIAM HOGARTH. Contain-ing more than 150 Illustrations reproduced in Permanent Photography, from very fine impressions of the Original En-gravings, collected by the late Mr. Dyce. With an Essay on his Genius and Character, by Charles Lamb, and a Description of each Picture. 2 vols. 4to. Handsomely bound. £5 5s.

PICTURESQUE VIEWS IN ENGLAND AND WALES. By J. M.W. Turner, R.A. With Descriptive Notices.

The "Picturesque Views in England and Wales," which were published some fifty years ago under Turner's especial care, are, next to the "Liber Studiorum," the most important series of engraved works which that great artist has bequeathed to us.

This magnificent work has long been out of print, and when-ever a copy is offered for sale it is eagerly purchased at an advance on its original price. It comprises Ninety-six Illustra-tions, which have now been reproduced by the Autotype process of permanent photography, from early India proof impres-sions of the original edition. The work is now bound in three volumes, the size being the same as the imperial quarto edition, published at Twenty-four Guineas. The price of the present edition is £6 6s.

For the convenience of those not wishing to purchase the whole work, the volumes are sold separately, as under :—

Landscapes	. 40 plates,	£2 12s. 6d.
Castles and Abbeys	. 32 ,,	£2 2s.
Coast Scenery	. 24 ,,	£1 11s. 6d.

TURNER'S CELEBRATED LANDSCAPES. This volume contains Sixteen Autotype Reproductions of the most important Works of J. M. W. Turner, R.A. With a Memoir and Descriptions. *New Edition.* Imperial 4to. Handsomely bound. £2 2s.

THE REMBRANDT GALLERY. Containing Thirty of the most important of Rembrandt's Etchings, reproduced by Photography, *the same size as the originals*, from the celebrated Collection in the British Museum. Selected by G. W. Reid, Keeper of the Prints and Drawings. With Notes on the Etchings by H. Noel Humphreys, Author of "Masterpieces of the Early Painters," &c. Imp. quarto. £3 3s.

THE RAFFAELLE GALLERY. A Series of Permanent Reproductions in Autotype of Engravings of the most celebrated Works of Raffaelle Sanzio d'Urbino. With Descriptions, &c. Imp. quarto. £2 2s.

THE LANDSEER GALLERY. A Series of Autotype Reproductions of Engravings of the celebrated Early Paintings of Sir Edwin Landseer, R. A. Imp. quarto. £2 2s.

MEMOIRS OF SIR EDWIN LANDSEER. Being a New Edition of "The Early Works of Sir Edwin Landseer," Revised and Enlarged by F. G. Stephens. With 24 Illustrations in Photography. Imperial 8vo. Handsomely bound. £1 5s.

MOUNTAIN, LOCH, AND GLEN. Illustrating "Our Life in the Highlands." From Paintings executed expressly for this Work, by Joseph Adam. With an Essay on the Highlands and Highlanders, by the late Rev. Norman Macleod, one of Her Majesty's Chaplains. *Cheaper edition.* Folio. Handsomely bound. £4 4s. With smaller Photographs. £3 3s. Smaller Edition. Imp. 4to. Autotype. £2 2s.

MOUNTAIN AND LAKES OF SWITZERLAND and Italy. Sixty-four Picturesque Views in Chromo-lithograph, after Drawings taken from the Original Sketches by C. C. Pyne. With a Map of Routes and Descriptive Notes, by the Rev. Jerome J. Mercier. *Second Edition.* Crown 4to. £2 2s.

THE RIVIERA. PEN AND PENCIL SKETCHES from Cannes to Genoa. By the late H. Alford, Dean of Canterbury. With Twelve Chromo-Lithographic Illustrations and numerous Woodcuts, from Drawings by the Author. Imp. 8vo. 21s.

THE ART OF SKETCHING FROM NATURE. By Philip H. Delamotte, Professor of Drawing at King's College, London. Illustrated with Twenty-four Woodcuts and Twenty-five coloured plates, arranged progressively, from Water-Colour Drawings by Prout, E. W. Cooke, R. A., Girtin, Varley, De Wint, Birket Foster, G. Thomas, and the Author. Imperial 4to. £3 3s.

A COLOURED EDITION OF MR. HENRY SHAW'S
Art of Illumination, as practised during the Middle Ages.
With a Description of the Metals, Pigments, &c., employed at
Different Periods. Imp. 8vo. Very handsomely bound. £1 1s.

ART AND SONG. Illustrated by Painters and Poets.
Containing thirty-one Steel Engravings from Drawings by
J. M. W. Turner, R. A., David Roberts, R. A., John Martin,
T. Stothard, R. A., F. Goodall, R. A., T. Uwins, R. A., W.
Collins, R. A., &c. Edited by Robert Bell. Demy 4to.
Walnut binding, £2 2s.

—— An Edition in folio. Cloth gilt. £2 2s.

A CATALOGUE OF WEDGWOOD'S MANUFAC-
TURES. With Illustrations. Half-bound. Edited by Eliza
Meteyard. 8vo. 10s. 6d.
This volume is reprinted at the Chiswick Press from the Ori-
ginal, which is very rare, and has been recently sold for £8.

HANDBOOK FOR THE COLLECTORS OF
WEDGWOOD WARE. By Eliza Meteyard. *In the Press.*

Art and Design.

A CONCISE HISTORY OF PAINTING FOR STU-
DENTS AND GENERAL READERS. By Mrs. Charles
Heaton, author of "The History of the Life of Albrecht Dürer
of Nürnberg," &c. In 8vo. With Illustrations in Permanent
Photography. 15s.
This volume does not pretend to give an exhaustive treatment of the subject, but, with the simple object of helping others
to enjoy good art, the author furnishes in an interesting form a
comprehensive and connected survey of the progress of painting
from the earliest historical periods to modern times.

LECTURES AND LESSONS ON ART. Being an
Introduction to a Practical and Comprehensive Scheme. By
F. W. Moody, Instructor in Decorative Art at South Kensington Museum. With Diagrams to Illustrate Composition and
other matters. Demy 8vo. 10s. 6d.
"How thoroughly Mr. Moody makes nature the inexhaustible fountain
of all artistic inspiration can be best judged of by an examination of his
illustrative plates. Nearly all of these are reproductions of natural form,
and more particularly of the human figure. . . . . We are sure that
this is the right direction in which we have to look for real progress in
English art."—*Saturday Review.*

SOCIETY OF ARTS DRAWING BOOK. A Draw-
ing Book, prepared at the request of the Society of Arts, by
John Bell, Sculptor. Oblong 4to. Half-bound. 3*s.* 6*d.*
It contains more than One Hundred and Fifty Illustrations,
from the Flat, from the Round, the Human Figure, Geometrical
Designs, Landscape, &c. &c., and One Hundred Pages of easy
instructions for the Student, adapted for general, household, or
even self-education, without the aid of a professional teacher.

DRAWING COPIES. By P. H. Delamotte, Professor
of Drawing at King's College, London. Ninety-six Original
Sketches in Architecture, Trees, Figures, Foregrounds, Landscapes, Boats, and Sea-pieces. Royal 8vo. Oblong, half-
bound. 12*s.*

RAFFAELLE: ORIGINAL STUDIES in the Univer-
sity Galleries, Oxford. Etched in facsimile by Joseph Fisher,
with Introduction. 4to. Cloth gilt. £1 11*s.* 6*d.*

MICHAEL ANGELO: ORIGINAL STUDIES,
in the University Galleries, Oxford. Etched in facsimile by
Joseph Fisher, with Introduction. 4to. Cloth gilt. 21*s.*

THE PASSION OF OUR LORD JESUS CHRIST,
Pourtrayed by Albert Dürer. Edited by Henry Cole, C.B.
Small 4to. 12*s.* 6*d.* cloth ; antique calf, 21*s.*

By John Flaxman, R.A., Sculptor.

FLAXMAN'S CLASSICAL COMPOSITIONS, com-
prising the Outline Illustrations to Homer's "Iliad" and
"Odyssey," the "Tragedies of Æschylus," "The Theogony
and Works and Days of Hesiod," Engraved by Piroli, of
Rome, and William Blake. Imp. 4to. Half-bound morocco.
£4 14*s.* 6*d.* The four parts separately, 21*s.* each.

FLAXMAN'S ILLUSTRATIONS TO THE DIVINE
POEM OF DANTE ALIGHIERI. With full description of
each composition from the translation of the Rev. Henry
Francis Cary. Imp. 4to. Handsomely half-bound in scarlet
morocco, £2 12*s.* 6*d.* Cloth, £2 2*s.* "The Inferno" alone,
cloth gilt, £1 1*s.*

FLAXMAN'S ACTS OF MERCY. Oblong folio.
Half-bound. 16*s.*

By George Cruikshank.

A COMPLETE CATALOGUE OF THE ENGRAVED
WORKS OF GEORGE CRUIKSHANK. Including Etchings on Steel, Copper, &c., and Woodcuts executed between
the years 1805 and 1870. Compiled by G. W. Reid, Keeper
of the Prints and Drawings in the British Museum. With a
large number of Illustrations chiefly from the Original Plates
and Blocks. In 3 vols. Royal 4to. £12 12s.
   As only 135 copies of this work have been printed and few
   copies only remain, early application is desirable.

THE TABLE BOOK. By John Oxenford, Horace
Mayhew, Gilbert A'Beckett, Mark Lemon, Shirley Brooks, W.
M. Thackeray, &c. Twelve Steel Plates and 116 Woodcuts
by George Cruikshank. Imperial 8vo. 16s.

THE OMNIBUS. 100 Illustrations on Steel and
Wood by George Cruikshank. Edited by Laman Blanchard.
Royal 8vo. 10s. 6d.

BOMBASTES FURIOSO. A Burlesque Tragic Opera,
in One Act, by William Barnes Rhodes. With Illustrations by
George Cruikshank. 3s. 6d.

THE TRUE LEGEND OF ST. DUNSTAN AND
THE DEVIL. Showing how a Horseshoe came to be a
Charm against Witchcraft. By Edward G. Flight, with Illustra-
tions by George Cruikshank. 3s. 6d.

PUNCH AND JUDY. The Dialogue of the Puppet
Show, an Account of its Origin, &c. New Edition. Post
8vo. With 24 Illustrations, designed and engraved by
George Cruikshank. 7s. 6d. Also, with Coloured Illustrations,
10s. 6d.

THE FAIRY LIBRARY, consisting of Hop o' My
Thumb, Puss in Boots, Cinderella, and Jack and the Bean-
stalk. With numerous fine Etchings by G. Cruikshank. 5s.

THE LOVING BALLAD of Lord Bateman. 1s. 6d.

See also Books by Mrs. Ewing, page 21 ; also pp. 34 and 38.

Guinea Gift Books.

In handsome bindings.

THE WORLD'S PICTURES. Photographs of Fifteen of some of the most celebrated Paintings in the World. Copied from the best Engravings. With an Introduction by C. C. Black, M.A. In one handsome demy 4to. volume. 21*s.*

THE SHEEPSHANKS GALLERY. Twenty Autotype Reproductions of the Pictures in the Sheepshanks Gallery in the South Kensington Museum. Demy 4to. Elegantly bound. 21*s.*

THE RUINS OF POMPEII. Photographic Views, with a History of the Destruction of the City, and Description of the most interesting Remains. By Thomas H. Dyer, LL.D. Demy 4to. 21*s.*

MASTERPIECES OF ENGLISH ART. Reproduced in Photography. With Memoirs of the Artists by C. Monkhouse. Demy 4to. 21*s.*

MASTERPIECES OF ITALIAN ART, Reproduced in Photography. With Memoirs. Demy 4to. 21*s.*

BIRKET FOSTER'S SUMMER SCENES. Fifteen Photographs from Drawings by Birket Foster, accompanied with appropriate Selections from the Poems of Clare, Bloomfield, Thomson, &c. 4to. Handsomely bound. 21*s.*

THE BOOK OF GEMS. Selections from the British Poets, illustrated with upwards of 150 Steel Engravings. Edited by S. C. Hall. 3 vols. Handsomely bound in walnut. 21*s.* each.
FIRST SERIES—CHAUCER TO DRYDEN.
SECOND SERIES—SWIFT TO BURNS.
THIRD SERIES—WORDSWORTH TO TENNYSON.

ADELAIDE ANNE PROCTER'S LEGENDS AND LYRICS. The Illustrated Edition. With Additional Poems, and an Introduction by Charles Dickens, a Portrait by Jeens, and 20 Illustrations by Eminent Artists. Fcap. 4to. Ornamental cloth. 21*s.*

HENRY WADSWORTH LONGFELLOW'S POETICAL WORKS. With nearly 250 Illustrations by Birket Foster, Tenniel, Godwin, Thomas, &c. In 1 vol. 21*s.*

MRS. GATTY'S PARABLES FROM NATURE; a
Handsomely ·Illustrated Edition ; with Notes on the Natural
History, and numerous full-page Illustrations by the most
eminent artists of the present day. Fcap. 4to.   21*s.*   Also in
2 volumes, 10*s.* 6*d.* each.

THE BOOK OF SUN-DIALS, collected and arranged
by Mrs. Alfred Gatty, author of "Parables from Nature."
With descriptions and mottoes of nearly 400 dials and 21 Litho-
graphic illustrations. Demy 4to.   21*s.*
    This work treats of the poetry and moral aspects of sun-dials,
as exhibited in their mottoes, under the varied treatment of
different countries and ages.

WASHINGTON IRVING'S SKETCH-BOOK. (The
Artist's Edition.)   Illustrated with a Portrait of the Author on
Steel, and 200 Exquisite Wood-Engravings from the Pencils of
the most celebrated American Artists. Crown 4to.   21*s.*

Gift Books, illustrated by Birket Foster and other Artists.
16mo. handsomely bound in malachite bindings.   9*s.*
each.

1. **THE WAYSIDE INN.** By Henry Wadsworth Longfellow.

2. **EVANGELINE.**   By Henry Wadsworth Longfellow.

3. **THE WHITE DOE OF RYLSTONE;** or, the Fate
of the Nortons.   By William Wordsworth.

4. **OLIVER GOLDSMITH'S POEMS.**

Books of Travel and Archæology.

ROME AND THE CAMPAGNA. A Historical and Topographical Description of the Site, Buildings, and Neighbourhood of ancient Rome. By the Rev. Robert Burn, late Fellow and Tutor of Trinity College, Cambridge. With eighty engravings by Jewitt, and numerous Maps and Plans. Demy 4to. £3 3s.

ANCIENT ATHENS; its History, Topography, and Remains. By Thomas Henry Dyer, LL.D., Author of "The History of the Kings of Rome." Super-royal 8vo. Illustrated, cloth. £1 5s.

This gives the result of the excavations to the present time, and of a recent careful examination of the localities by the Author. It is illustrated with plans and wood engravings taken from photographs.

THE HISTORY OF POMPEII; its Buildings and Antiquities. An account of the City, with a full description of the Remains and the Recent Excavations, and also an Itinerary for Visitors. Edited by T. H. Dyer, LL.D. Illustrated with nearly 300 Wood Engravings, a large Map, and a Plan of the Forum. *The Second Edition.* 8vo. 14s.

THE INVASION OF BRITAIN by Julius Cæsar. With replies to the remarks of the Astronomer Royal, and of the late Camden Professor of Ancient History at Oxford. By Thomas Lewin, M.A., Trinity College, Oxford, Barrister-at-law, F.S.A With Maps and Illustrations. Demy 8vo. 10s. 6d.

THE DESERT OF THE EXODUS. Journeys on Foot in the Wilderness of the Forty Years' Wanderings, undertaken in connection with the Ordnance Survey of Sinai and the Palestine Exploration Fund. By E. H. Palmer, M.A., Lord Almoner's Professor of Arabic and Fellow of St. John's College, Cambridge, Member of the Asiatic Society, and of the Société de Paris. With Maps, and numerous Illustrations from Photographs and Drawings taken on the spot by the Sinai Survey Expedition and C. F. Tyrwhitt Drake. 2 vols. 8vo. 28s.

"A work which the biblical student will highly prize for the strong light which it sheds upon a most important portion of Scripture history, but which cannot be read without interest and delight by every one who is capable of taking an intelligent interest in manners and customs widely removed from our own."—*Saturday Review.*

THE HISTORY OF EGYPT. From the Earliest Times till its Conquest by the Arabs, A.D. 640. By Samuel Sharpe. *New Edition, revised and enlarged.* 2 vols. Large post 8vo. With numerous Illustrations, Maps, &c. Cloth. 18s.

THE FOOTSTEPS OF OUR LORD AND HIS
APOSTLES IN PALESTINE, SYRIA, GREECE, AND
ITALY. By W. H. Bartlett. *Seventh edition*, with numerous
engravings. In one 4to. volume. Handsomely bound in walnut,
18*s.* Cloth gilt, 10*s.* 6*d.*

FORTY DAYS IN THE DESERT ON THE TRACK
OF THE ISRAELITES; or, a Journey from Cairo to Mount
Sinai and Petra. By W. H. Bartlett. 4to. With 25 steel en-
gravings. Handsome walnut binding, 18*s.* Cloth gilt, 10*s.* 6*d.*

THE NILE BOAT; or, Glimpses of the Land of Egypt.
By W. H. Bartlett. *New edition*, with 33 steel engravings. 4to.
Walnut, 18*s.* Cloth gilt, 10*s.* 6*d.*

THE BOAT AND THE CARAVAN. A Family Tour
through Egypt and Syria. *New and cheaper Edition.* Fcap.
8vo. 5*s.* 6*d.*

DOMESTIC LIFE IN PALESTINE. By M. E.
Rogers. *Second Edition.* Post 8vo. 10*s.* 6*d.*

LETTERS FROM INDIA AND KASHMIR. Anno-
tated with numerous fine engravings. Fcap. 4to. £1 11*s.* 6*d.*

COLONIAL ADVENTURES AND EXPERIENCES.
By George Carrington. Crown 8vo. 7*s.* 6*d.*

BEHIND THE SCENES IN RUSSIA. By George
Carrington, B.A., Oxford. Crown 8vo. 7*s.* 6*d.*

PREHISTORIC PHASES; or, Introductory Essays on
Prehistoric Archæology. By Hodder M. Westropp. Demy
8vo. 9*s.*

FLINT CHIPS; a Guide to Prehistoric Archæology as
illustrated by the Collection in the Blackmore Museum, Salis-
bury. By E. T. Stevens. Demy 8vo. With numerous Illustrations. 15*s.*

WEAPONS OF WAR. A History of Arms and Armour
from the Earliest Period to the Present Time. Translated from
the French of Auguste Demmin, by C. C. Black, M.A. With
nearly 2,000 Illustrations. 12*s.*

ENGRAVINGS OF UNEDITED OR RARE GREEK
COINS. With Descriptions. By General C. R. Fox. 4to.
Part I. Europe. Part II. Asia and Africa. 7*s.* 6*d.* each.

By Rev. C. W. King, M. A., Fellow of Trinity College, Cambridge.

EARLY CHRISTIAN NUMISMATICS, and other Antiquarian Tracts contributed to the "Archæological Journal" and other papers. 8vo. Illustrated. Cloth gilt. 18*s.*

ANTIQUE GEMS AND RINGS. *Second Edition,* greatly enlarged and improved, with more than 600 Illustrations. 2 vols. Super-royal 8vo. £2 2*s.*

THE NATURAL HISTORY of PRECIOUS STONES AND OF THE PRECIOUS METALS. *New Edition, revised.* Post 8vo. Illustrated. 6*s.*

THE NATURAL HISTORY OF GEMS OR DECORATIVE STONES. *New Edition, revised.* Post 8vo. Illustrated. 10*s. 6d.*

HANDBOOK OF ENGRAVED GEMS. Illustrated with numerous Plates. Crown 8vo. 10*s. 6d.*

This volume is intended as a manual for the use of the student and collector of engraved gems. It contains a history of the Glyptic Art, an account of the celebrated European cabinets and famous rings and signets, with artists' signatures, lists of ancient artists, &c. It is illustrated with nearly two hundred wood engravings.

THE GNOSTICS AND THEIR REMAINS. 8vo. 15*s.*

THE SCIENCE OF GEMS, JEWELS, COINS, AND MEDALS, ANCIENT AND MODERN. By Archibald Billing, M.D., A.M., F.R.S., &c. Illustrated with Photographs of 160 Ancient and Modern Specimens. Demy 8vo. 31*s. 6d.*

THE ARCHITECTURAL HISTORY OF GLASTONBURY ABBEY. By R. Willis, M.A., F.R.S., Jacksonian Professor, Cambridge. With Illustrations. 8vo. 7*s. 6d.*

A CHRONICLE OF THE CHURCH OF ST. MARTIN, LEICESTER, during the reigns of Henry VIII., Edward VI., Mary and Elizabeth. By Thomas North. Fcap. 4to. 21*s.*

THE ARCHITECTURAL HISTORY OF EXETER CATHEDRAL. By Rev. P. Freeman, Archdeacon and Canon of Exeter. Fcap. 4to. 8*s. 6d.*

THE ILLUSTRATED HANDBOOK FOR WEST-MINSTER ABBEY. By Felix Summerly. *6d.*

THE ABBEY OF ST. ALBAN. An Historical and Descriptive Account, by H. J. B. Nicholson, D.D., F.S.A. *New edition.* With an account of the Shrine of St. Alban, recently discovered, by W. J. Lawrance, Rector of St. Alban's. Demy 8vo. *1s. 6d.*

ANECDOTES OF HERALDRY; in which is set forth the Origin of the Armorial Bearings of many Families. With 100 Illustrations. By C. N. Elvin, M.A. Fcap. 8vo. *10s. 6d.*

A HANDBOOK OF MOTTOES borne by the Nobility, Gentry, Cities, Public Companies, &c. Translated and Illustrated with Notes and Quotations. By C. N. Elvin, M.A. *6s.*

Natural History.

SOWERBY'S ENGLISH BOTANY. Containing a Description and Life-size Drawing of every British Plant. Edited and brought up to the present standard of scientific knowledge, by T. Boswell Syme, LL.D., F.L.S., &c. With Popular Descriptions of the Uses, History, and Traditions of each Plant, by Mrs. Lankester, Author of "Wild Flowers Worth Notice," "The British Ferns," &c. The Figures by J. C. Sowerby, F.L.S., J. De C. Sowerby, F.L.S., and J. W. Salter, A.L.S., F.G.S. and John Edward Sowerby. Third edition, entirely revised, with descriptions of all the species by the editor.

	Bound in cloth.			Half morocco.			Morocco elegant.		
	£	s.	d.	£	s.	d.	£	s.	d.
Vol. 1. (Seven Parts) . .	1	18	0	2	2	0	2	8	6
Vol. 2. ditto . .	1	18	0	2	2	0	2	8	6
Vol. 3. (Eight Parts) . .	2	3	0	2	7	0	2	13	6
Vol. 4. (Nine Parts) . .	2	8	0	2	12	0	2	18	6
Vol. 5. (Eight Parts) . .	2	3	0	2	7	0	2	13	6
Vol. 6. (Seven Parts) . .	1	18	0	2	2	0	2	8	6
Vol. 7. ditto . .	1	18	0	2	2	0	2	8	6
Vol. 8. (Ten Parts) . .	2	13	0	2	17	0	3	3	6
Vol. 9. (Seven Parts) . .	1	18	0	2	2	0	2	8	6
Vol. 10. ditto . .	1	18	0	2	2	0	2	8	6
Vol. 11. (Six Parts) . .	1	13	0	1	17	0	2	3	6

Or, the Eleven Volumes, £22 8s. in cloth ; £24 12s. in half morocco, and £28 3s. 6d. whole morocco.

SOWERBY'S FERN AND FERN-ALLIES OF GREAT BRITAIN. Illustrated with 80 plates by John E. Sowerby. The descriptions, synonyms, &c., by Charles Johnson. 8vo., 10s. 6d. Royal paper, coloured plates, 25s.

THE BOTANIST'S POCKET-BOOK. By W. R. Hayward. Containing, arranged in a tabulated form, the chief characteristics of British Plants. Fcap. 8vo., flexible binding for the pocket, 4s. 6d.

This volume is intended as a handy Pocket Companion for the Botanist in the field, and will enable him to identify on the spot the plants he may meet with in his researches. Besides the characteristics of species and varieties, it contains the Botanical name, Common name, Soil or Situation, Colour, Growth, and time of Flowering of every plant, arranged under its own order.

Library of Natural History.

"We have watched with much pleasure the gradual publication of an extensive and valuable series of works on the Natural History of Birds, Fishes, Butterflies, and Plants. Each volume is elegantly printed in royal 8vo., and illustrated with a very large number of well-executed engravings, printed in colours. . . . . The authors and artists to whom have been entrusted the text and illustration of these original works are men of note and capacity in their several departments ; and it is with pleasure we invite the attention of our readers to these handsome books. . . They form a complete library of reference on the several subjects to which they are devoted, and nothing more complete in their way has lately appeared. Although forming a library of natural history, each subject is in fact complete in itself."—*The Bookseller.*

BREE'S HISTORY OF THE BIRDS OF EUROPE,
not observed in the British Isles ; and their eggs. With 238 beautifully coloured Plates. 4 vols. *New edition preparing.*

COUCH'S HISTORY OF THE FISHES OF THE BRITISH ISLANDS. With 256 carefully coloured Plates. 4 vols. *New edition preparing.*

GATTY'S BRITISH SEAWEEDS. Drawn from Professor Harvey's "Phycologia Britannica," with Descriptions in popular language by Mrs. Alfred Gatty. 2 vols. £2 10s.
This volume contains drawings of the British Seaweeds in 384 figures coloured after nature, with descriptions of each, including all the newly-discovered species ; an Introduction, an Amateur's Synopsis, Rules for preserving and laying out Seaweeds, and the Order of their arrangement in the Herbarium.

HIBBERD'S NEW AND RARE BEAUTIFUL-LEAVED PLANTS, containing fifty-four coloured Plates, executed expressly for this work. 1 vol. 25s.

LOWE'S NATURAL HISTORY OF BRITISH AND EXOTIC FERNS. With 479 finely coloured Plates. 8 vols. £6 6s.

LOWE'S OUR NATIVE FERNS; a History of the British Species and their Varieties. Containing descriptions of Fifty Species and 1,300 Varieties. With 79 coloured Plates and 909 Engravings. 2 vols. £2 2s.

LOWE'S NATURAL HISTORY OF NEW AND RARE FERNS. Containing Species and Varieties not included in "Ferns, British and Exotic." 72 coloured Plates and Woodcuts. 1 vol. 21s.

LOWE'S NATURAL HISTORY OF BRITISH GRASSES. With 74 finely coloured Plates. 1 vol. £1 1s.

LOWE'S BEAUTIFUL-LEAVED PLANTS, being a Description of the most beautiful-leaved Plants in cultivation. With 60 coloured illustrations. £1 1s.

MAUND'S BOTANIC GARDEN. *New edition, revised.* In six volumes. Containing 240 coloured plates. *In the Press.*

MORRIS'S HISTORY OF BRITISH BIRDS. 365 coloured Plates. 6 vols. £6 6s.

MORRIS'S NESTS AND EGGS OF BRITISH BIRDS. With 233 beautifully coloured Engravings. 3 vols. £3 3s.

MORRIS'S BRITISH BUTTERFLIES. 72 beautifully coloured Plates. £1 1s.

MORRIS'S BRITISH MOTHS. With Coloured Illustrations of nearly 2,000 Specimens. 4 vols. £6 6s.

TRIPP'S BRITISH MOSSES. 39 Coloured Plates. Containing a figure of every Species. *New edition.* 2 vols. £2 10s.

WOOSTER'S ALPINE PLANTS. Figures and Descriptions of the most striking and beautiful of the Alpine Flowers. 54 coloured plates. 25s.
A second volume in preparation.

MY GARDEN; its Plan and Culture. Together with a General Description of its Geology, Botany, and Natural History. By Alfred Smee, F.R.S. Illustrated with more than 1,300 Engravings on Wood. *Second edition,* revised, imperial 8vo. 21s.

The purpose of this work is to describe the author's garden at Wallington in Surrey, and to serve as an exact and practical guide to amateurs in every branch of Horticulture practised in general gardening. Illustrations are given of garden scenes as specimens of Landscape Gardening ; of Tools, Frames, and Glass Houses ; Vegetables, Fruits, Flowers, Weeds and Wild Flowers ; Fungi, Ferns, Trees, Shrubs, Animals, Birds, Insects ; also modes of Grafting Trees, &c. It will be found very useful as a book of reference to all persons partial to Horticulture or having a love of nature.

B

MRS. LOUDON'S FIRST BOOK OF BOTANY.
Revised and enlarged by David Wooster. With numerous
illustrations. Fcap. 8vo. Cloth. 2s. 6d.

THE COTTAGE GARDENER'S DICTIONARY.
With a Supplement, containing all the new plants and varieties
now cultivated. Edited by George W. Johnson. Post 8vo.
Cloth. 6s. 6d.

On Medicine.

THE FIRST PRINCIPLES OF MEDICINE. By
Archibald Billing, M.D., F.R.S. Thoroughly revised and
enlarged, and brought down to the present state of medical
science. *Sixth edition.* 8vo. 18s.

> "We know of no book which contains within the same space so much
> valuable information, the result, not of fanciful theory, nor of idle hypo-
> thesis, but of close persevering clinical observation, accompanied with
> much soundness of judgment and extraordinary clinical tact."—*Medico-
> Chirurgical Review.*

MEDICINE AND PSYCHOLOGY. An Address to
the Hunterian Society. By Dr. De B. Hovell, F.R.C.S.
Crown 8vo. Cloth. 3s. 6d.

LATHAM (P. W.) On the Symptomatic Treatment of
Cholera with especial reference to the importance of the Intes-
tinal Lesion. By Dr. Felix von Niemeyer, Professor of
Pathology and Therapeutics, &c., in the University of Tübingen.
Translated from the German by P. W. Latham, M.A., M.D.,
Fellow of the Royal College of Physicians, London, late Fellow
of Downing College, Cambridge. Small 8vo. 2s. 6d.

—— On Nervous or Sick Headache: its Varieties and
Treatment : two Lectures Delivered at Addenbrooke's Hospital,
Cambridge. Post 8vo. Pp. 72. Cloth. 3s.

**ON THE DISEASE OF THE RIGHT SIDE OF
THE HEART.** By Dr. Daldy, L.R.C.P. Post 8vo. 3s. 6d.

Books for Young People.

Mrs. Alfred Gatty's Popular Works.

PARABLES FROM NATURE. With Notes on the Natural History ; and numerous large Illustrations by eminent artists. 4to. cloth gilt. 21*s.* Also in two volumes, 10*s. 6d.*

PARABLES FROM NATURE. 16mo. with Illustrations. First series, *Sixteenth Edition*, 1*s. 6d.* Second Series, *Tenth Edition*, 2*s.* The two Series in one volume, 3*s. 6d.* Third Series, *Sixth Edition*, 2*s.* Fourth Series, *Fourth Edition*, 2*s.* The two Series in one volume, 4*s.* Fifth Series, 2*s.*

SELECT PARABLES FROM NATURE, for the use of Schools. Fcap. 1*s. 6d.*
Besides being reprinted in America, selections from Mrs. Gatty's Parables have been translated and published in the German, French, Italian, Russian, Danish, and Swedish languages.

WORLDS NOT REALIZED. 16mo. *Fourth Edition.* 2*s.*

PROVERBS ILLUSTRATED. 16mo. With Illustrations. *Fourth Edition.* 2*s.*

A BOOK OF EMBLEMS. Drawn on Wood by F. Gilbert. With Verbal Illustrations and an Introduction by Mrs. Alfred Gatty. Imp. 16mo. 4*s. 6d.*

WAIFS AND STRAYS OF NATURAL HISTORY. With Coloured Frontispiéce and Woodcuts. Fcap. 3*s. 6d.*

THE POOR INCUMBENT. Fcap. 8vo. Sewed, 1*s.* Cloth, 1*s. 6d.*

AUNT SALLY'S LIFE. With Six Illustrations by G. Thomas. Square 16mo. 3*s. 6d.* (Revised reprint from "Aunt Judy's Letters.")

THE MOTHER'S BOOK OF POETRY. Selected
and Arranged by the late Mrs. Alfred Gatty. Crown 8vo.
With Illustrations, elegantly bound. 7s. 6d.

A BIT OF BREAD. By Jean Macé. Translated from
the French by Mrs. Alfred Gatty. 2 vols. fcap. 8vo. Vol. I.
4s. 6d. ; Vol. II. 3s. 6d.

A PORTRAIT OF MRS. GATTY has recently been engraved. A
few copies on India paper may be had, price 5s.

The Uniform Edition of Mrs. Alfred Gatty's Works.
Fcap. 8vo. size.

PARABLES FROM NATURE. Two Volumes. With
Portrait of Mrs. Alfred Gatty. Price 3s. 6d. each.

WORLDS NOT REALIZED, AND PROVERBS
Illustrated. In one vol. 3s. 6d.

DOMESTIC PICTURES AND TALES. With Six Illustrations. 3s. 6d.

AUNT JUDY'S TALES. Illustrated by Clara S. Lane.
Fifth Edition. 3s. 6d.

AUNT JUDY'S LETTERS : a Sequel to "Aunt
Judy's Tales." Illustrated by Clara S. Lane. *New Edition.*
3s. 6d.

THE HUMAN FACE DIVINE, and other Tales.
With Illustrations by C. S. Lane. *Second Edition.* 3s. 6d.

THE FAIRY GODMOTHERS, and other Tales. With
Frontispiece. *Fifth Edition.* 2s. 6d.

THE HUNDREDTH BIRTHDAY, and other Tales.
With Illustrations by Phiz. *New Edition, revised.* 3s. 6d.

MRS. ALFRED GATTY'S PRESENTATION BOX
for YOUNG PEOPLE, containing the above volumes, all
beautifully printed, neatly bound, and enclosed in a cloth box.
31s. 6d.

AUNT JUDY'S MAGAZINE VOLUMES. Edited by Mrs. Alfred Gatty. Imperial 16mo., handsomely bound. Containing Stories, Songs, Music, Papers on Natural History, Fairy Tales, etc., etc. With numerous Illustrations. Christmas volumes, 1872, and 1873 (double volumes), 7s. 6d. each. The early volumes may be had at a reduced price.

⁎ The Monthly Sixpenny Part for November, 1873, begins the Christmas Volume for 1874, under the Editorship of Mrs. Gatty's daughters, H. K. F. Gatty, and J. H. Ewing. All the parts are kept in print. Price 6d. each.

By Mrs. Ewing (J. H. Gatty).

LOB LIE-BY-THE-FIRE; or, the Luck of Lingborough : and other Tales. Illustrated by George Cruikshank. Imperial 16mo. 5s.

THE BROWNIES, and other Tales. Illustrated by George Cruikshank. *Second edition.* Crown 8vo. 5s.

A FLAT IRON FOR A FARTHING. With Illustrations by H. Paterson. Fcap. 8vo. 5s.

MRS. OVERTHEWAY'S REMEMBRANCES. Illustrated with Nine full-page Engravings on Wood by Pasquier, and a Frontispiece by Wolf. Post 8vo. Cloth and gold. *Second Edition.* 4s.

MELCHIOR'S DREAM, and other Tales. Illustrated. *Second edition.* Fcap. 8vo. 3s. 6d.

AUNT JUDY'S SONG-BOOK FOR CHILDREN : Containing Twenty-four Popular Songs, etc. By Alfred Scott Gatty, Composer of "O! Fair Dove, O! Fond Dove," etc. Fcap. 4to. 4s. 6d.

ROBINSON CRUSOE. With a Biographical Account of Defoe. Illustrated with Seventy Wood Engravings, chiefly after designs by Harvey ; and Twelve Engravings on Steel after Stothard. Post 8vo. 5s.

Captain Marryat's Books for Boys.

POOR JACK. With Forty-seven Illustrations after
Designs by Clarkson Stanfield, R.A. *Twenty-second Edition.*
8vo. Cloth gilt. 6s. Also *Cheap edition.* Fcap. 8vo. with
Frontispiece. 3s. 6d.

THE MISSION ; OR, SCENES IN AFRICA. With
Illustrations by John Gilbert. 5s. ; or in ornamental binding,
gilt edges, 5s. 6d.

THE SETTLERS IN CANADA. With Illustrations
by Gilbert and Dalziel. 5s. ; or in ornamental binding, gilt
edges, 5s. 6d.

THE PRIVATEER'S MAN. Adventures by Sea and
Land in Civil and Savage Life, One Hundred Years Ago.
Illustrated with Eight Steel Engravings, 5s. ; or in ornamental
binding, gilt edges, 5s. 6d.

MASTERMAN READY ; or, the Wreck of the Pacific.
Embellished with Ninety-three Engravings on Wood. 5s. ; or
in ornamental binding, 5s. 6d.

THE PIRATE AND THREE CUTTERS. Illus-
trated with Twenty Steel Engravings from Drawings by
Clarkson Stanfield, R.A. With a Memoir of the Author.
5s. ; or in ornamental binding, gilt edges, 5s. 6d.

A BOY'S LOCKER. A Uniform Edition of the above
in 12 volumes, enclosed in a compact cloth box. 25s.

HANS CHRISTIAN ANDERSEN'S FAIRY TALES
AND SKETCHES. Translated by C. C. Peachey, H. Ward,
A. Plesner, &c. With 104 Illustrations by Otto Speckter and
others. 6s.
   This volume contains several tales that are in no other
Edition published in this country, and with the following
volume it forms the most complete English edition.

HANS CHRISTIAN ANDERSEN'S TALES FOR
CHILDREN. With 48 Full-page Illustrations by Wehnert,
and 57 Small Engravings on Wood by W. Thomas. A *New
Edition.* Very handsomely bound. 6s.

THE LATER TALES OF HANS CHRISTIAN AN-
DERSEN. Translated from the Danish by Augusta Plesner and
H. Ward. With Illustrations by Otto Speckter, A. W. Cooper,
and other Artists. Handsomely bound in cloth gilt. 3s. 6d.

GRIMM'S GAMMER GRETHEL ; or, German Fairy
Tales and Popular Stories. Translated by Edgar Taylor.
Numerous Woodcuts, after G. Cruikshank's designs. Post
8vo. 3s. 6d.

MRS. LOUDON. The Entertaining Naturalist. Containing Descriptions and Anecdotes of more than Five Hundred
Animals, &c. ; with nearly 500 Illustrations. With an Introduction on the Classification of Animals. Carefully revised and
enlarged by W. S. Dallas, F.L.S. Post 8vo. 7s.

ADAMS' SMALLER BRITISH BIRDS. With descriptions of their Nests, Eggs, Habits, &c. Numerous
coloured Illustrations. 15s. 　　　　　*Just published.*

RENNIE. Insect Architecture. By James Rennie.
- Edited and Enlarged by the Rev. J. G. Wood, Author of
" Homes Without Hands." Post 8vo. With nearly 200 Illus-
trations. 5s.

GLIMPSES INTO PET-LAND. By the Rev. J. G.
Wood, M.A., F.L.S. &c. With Frontispiece by Crane. Fcap.
8vo. 3s. 6d.

FRIENDS IN FUR AND FEATHERS. By Gwyn-
fryn (one of the contributors to "Aunt Judy's Magazine").
Illustrated with Eight Full-page Engravings by F. W. Keyl
and other artists. *Third edition.* Handsomely bound. 4s.

" We have already characterized some other book as the best cat-and-
dog book of the season. We said so because we had not seen the present
little book, which is delightful. It is written on an artistic principle, consisting of actual biographies of certain elephants, squirrels, blackbirds, and
what not, who lived in the flesh ; and we only wish that human biographies
were always as entertaining and instructive."—*Saturday Review.*

ANECDOTES OF DOGS. By Edward Jesse. With
Illustrations by W. Harvey, W. P. Smith, Bewick, Seymour,
Radcliffe, and other Artists. Post 8vo. Cloth. 5s. With the
addition of Thirty-four Steel Engravings after Cooper, Landseer, &c. 7s. 6d.

THE NATURAL HISTORY OF SELBORNE. By
Gilbert White. Edited by Jesse. Illustrated with Forty Engravings. Post 8vo. 5s. ; or with the Plates Coloured, 7s. 6d.

WALKER'S MANLY EXERCISES. Containing Rowing, Sailing, Riding, Driving, Skating, Running, and other
manly sports. The whole carefully revised or written by
"Craven," with upwards of Forty Steel Plates and Numerous
Woodcuts. Post 8vo. 5s.

THE YOUNG SPORTSMAN'S MANUAL; or, Recreations in Shooting, with some account of the Game found in the British Islands, and Practical Directions for the Management of Dog and Gun. By "Craven." Illustrated. Post 8vo. 5*s.*

CRICKET. By Thomsonby. Containing—Its Origin and Laws, Batting, Fielding, Management of Matches and Cricketers in Council. 2*s.* 6*d.*

OXFORD AND CAMBRIDGE BOAT RACES. A Chronicle of the Contests on the Thames in which University Crews have borne a part, from A.D. 1829 to A.D. 1869. Compiled from the University Club Books and other Contemporary and Authentic Records, with Maps of the Racing Courses, Index of Names, and an Introduction on Rowing and its value as an Art and Recreation. By W. F. Macmichael, B.A., Downing College, Secretary C. U. B. C. Fcap. 8vo. 6*s.*

OARS AND SCULLS. By W. B. Woodgate, B.A., Brasenose College, Oxford, Amateur Champion of the Thames.
In the Press.

CHARADES, ENIGMAS, AND RIDDLES. Collected by a Cantab. *Fifth edition, enlarged.* Illustrated. Fcap. 8vo. 1*s.*

KATIE; or the Simple Heart. By D. Richmond, Author of "Annie Maitland." Illustrated by M. I. Booth. *Second Edition.* Crown 8vo. 6*s.*

OUVRY (Miss). Arnold Delahaize; or, the Huguenot Pastor. By Francisca Ingram Ouvry. With a Frontispiece. Fcap. 8vo. 2*s.* 6*d.*

—— Henri de Rohan. By the Author of "Arnold Delahaize." With 4 Illustrations. Post 8vo. 5*s.*

—— Hubert Montreuil; or, the Huguenot and the Dragoon. By the Author of "Arnold Delahaize." With Five Illustrations in Photography. Imperial 16mo. 8*s.*

MARIE; or, Glimpses of Life in France. Cr. 8vo. 6*s.*

MARIETTE; or, Further Glimpses of Life in France. A Sequel to "Marie." Crown 8vo. Cloth. 7*s.* 6*d.*

SYDONIE'S DOWRY. By the Author of "Denise" and "Mademoiselle Mori." Crown 8vo. 6*s.*

TALES, OLD AND NEW. By the Author of "Mademoiselle Mori." Post 8vo. 6s.

GREGORY HAWKSHAW. By George Carrington, B.A., Author of "Colonial Adventures." Post 8vo. 7s. 6d.

NINETY-THREE; or, the Story of the French Revolution, from the Recollections of my French Tutor. By John W. Lyndon. Crown 8vo. 7s. 6d.

STRICKLAND. Lives of the Queens of England from the Norman Conquest. By Agnes Strickland, author of " The Life of Mary, Queen of Scots." An Abridged Edition, with Portrait of Matilda of Flanders. In one Volume. Crown 8vo. Cloth. 6s. 6d.

NOBLE DEEDS OF WOMAN; or, Examples of Female Courage and Virtue. By Elizabeth Starling. *Seventh Edition.* Illustrated with Fourteen Steel Engravings. Post 8vo. 5s.

WOODROOFFE (MRS.) Cottage Dialogues. By the late Mrs. Woodrooffe. *New Edition.* 12mo. 4s. 6d.

—— Michael Kemp, the Happy Farmer's Lad. *Eighth Edition.* 12mo. 4s.

—— Sequel to Michael Kemp. *New Edition.* 12mo. 6s. 6d.

UNDINE. A Tale from the German. By the Hon. C. L. Lyttelton. With Illustrations. Fcap. Cloth. 1s.

THE WHITE LADY. A Tale from the German. By the Hon. C. L. Lyttelton. With Illustrations. Fcap. Cloth. 1s.

TWILIGHT AND DAWN. By the Author of "The Four Messengers." 16mo. 4s.

ECHOES. Stories for Children. By the Author of "Four Messengers." With a Frontispiece by E. J. Poynter, R.A. Royal 16mo. 3s. 6d.

KIRSTIN'S ADVENTURES. A Story of Jutland Life. By the Author of " Casimir the Little Exile," and translator of " Andersen's Fairy Tales," &c., with Illustrations. Crown 8vo. 6s.

Mrs. O'Reilly's " Daisy Library."

GILES'S MINORITY ; or, Scenes at the Red House.
With Eight Illustrations. 3*s.* 6*d.*　　　　*Just published.*

DOLL WORLD ; or, Play and Earnest. A Study from
Real Life. With Eight Illustrations by C. A. Saltmarsh. 3*s.* 6*d.*

DEBORAH'S DRAWER.　With Nine Illustrations
by H. Paterson. 3*s.* 6*d.*

DAISY'S COMPANIONS; or, Scenes from Child Life.
A Story for little Girls. With Eight Illustrations by A. A.
Hunt. 3*s.* 6*d.*

The Four Volumes, in a neat Cloth Box, 14*s.*

LITTLE MAGGIE AND HER BROTHER.　By
Mrs. G. Hooper. Fcap. Cloth. 1*s.*

THE LOST LEGENDS OF THE NURSERY SONGS.
By Mary Senior Clark. With Sixteen full-page Illustrations.
Imp. 16mo. 5*s.*

LITTLE PLAYS FOR LITTLE PEOPLE; with
Hints for Drawing-room Performances. By Mrs. Chisholm,
Author of " Rana, the Story of a Frog." 16mo. With Illus-
trations. 2*s.* 6*d.*

GUESSING STORIES; or, the Surprising Adventures
of the Man with the Extra Pair of Eyes. A Book for Young
People. By Rev. P. Freeman, Archdeacon. *Second Edition.*
Super-royal 16mo. Cloth. Gilt edges. 2*s.* 6*d.*

By Mrs. Harriet Beecher Stowe.

QUEER LITTLE PEOPLE. Fcap. Boards. 1*s.*;
cloth, 1*s.* 6*d.*

LITTLE FOXES ; or, the Little Failings that mar
Domestic Happiness. Fcap. Boards, 1*s.* ; cloth, 1*s.* 6*d.*
" Wise words, wisely, quaintly, and racily spoken."—*Nonconformist.*

CHIMNEY CORNER. Fcap. Boards, 1*s.* ; cloth,
1*s.* 6*d.*

By the Rev. J. Erskine Clarke, of Derby.

HEART MUSIC, for the Hearth-Ring; the Street-Walk; the Country Stroll; the Work-Hours; the Rest-Day; the Trouble-time. *New Edition.* 1s.

CHILDREN AT CHURCH. Twelve Simple Sermons. *New Edition.* 2 vols. 1s. each; together, in one vol. cloth gilt, 2s. 6d.

THE GIANT'S ARROWS. A Book for the Children of Working People. 16mo. 6d.

CHURCH STORIES. Edited by the Rev. J. E. Clarke. Crown 8vo.

SCRIPTURE PARABLES. 16 Illustrations. 1s.

BIBLE MIRACLES. 16 Illustrations. 1s.

NEW TESTAMENT HISTORY, in Simple Language. By Rev. J. G. Wood, M.A. 16 Illustrations. 1s.

OLD TESTAMENT HISTORY, in Simple Language. By the Rev. J. G. Wood, M.A. 17 Illustrations. 1s.

THE LIFE OF JOSEPH. 16 Illustrations. 1s.

THE STORY OF BUNYAN'S PILGRIM'S PROGRESS. 16 Illustrations. 1s.

THE PILGRIM'S PROGRESS. By John Bunyan. With 281 Engravings from Designs by William Harvey. Post 8vo. 3s. 6d.

A HISTORY OF GOD'S CHURCH OF OLD, for Children. By Mrs. Richard R. Armstrong, Author of "A History of Our Lord's Life on Earth." 16mo. 3s. 6d.

THE HISTORY OF OUR LORD'S LIFE ON EARTH. From the Four Gospels. For the use of Children. By E.S.A. Edited by the Rev. W. E. Heygate, M.A., Rector of Brighstone. Royal 16mo. 3s.

THE LIFE OF CHRISTOPHER COLUMBUS, in
Short Words. By Sarah Crompton. Super-royal 16mo.
2*s*. 6*d*. Also an Edition for Schools, 1*s*.

NURSERY TALES. By Mrs. Motherly. With Illustrations by C. S. Lane. Imperial 16mo. 2*s*. 6*d*. Coloured,
gilt edges, 3*s*. 6*d*.

BOOKS FOR YOUNG READERS, in Words of One
and Two Syllables. In Eight Parts. Sold separately. 1*s*. each,
neatly bound in cloth ; or plainly bound, 8*d*. each.

A POETRY BOOK FOR CHILDREN. Illustrated
with Thirty-seven highly-finished Engravings, by C. W. Cope,
R.A., Helmsley, Palmer, Skill, Thomas, and H. Weir. *New
edition.* Crown 8vo. 2*s*. 6*d*. A School Edition, cloth
limp, 1*s*.

GILES WITHERNE ; or, the Reward of Disobedience.
A Village Tale for the Young. By the Rev. J. P. Parkinson,
D.C.L. *Sixth edition.* Illustrated by the Rev. F. W. Mann.
Super-royal 16mo. 1*s*.

OLD NURSERY RHYMES AND CHIMES. Col-
lected and arranged by a Peal of Bells. Fcap. 4to. Ornamental binding. 2*s*. 6*d*.

NURSERY CAROLS, by the Rev. Dr. Monsell, Rector
of St. Nicholas, Guildford, with upwards of 100 Illustrations
by Ludwig Richter and Oscar Pletsch. Imperial 16mo. 3*s*. 6*d*.

English Literature and Poetry.

SHAKESPEARE. Shakespeare's Plays and Poems. With Notes and Life by Charles Knight, and 40 engravings on wood by Harvey. Royal 8vo. Cloth, 10s. 6d.

—— Shakespeare's Plays and Poems (Valpy's cabinet Pictorial Edition), with Glossarial Notes, Digests, etc., and 171 outline plates. 15 vols. Fcap. 8vo. £2 5s.

—— The Pocket Volume Shakespeare. Comprising all his Plays and Poems. Carefully Edited from the First Folio Edition by Thomas Keightley, printed at the Chiswick Press, neatly bound in 13 vols. royal 32mo. and enclosed in a cloth box, price 21s.

—— Charles Kemble's Shakespeare Readings. A Selection of Sixteen Plays as read by him before her Majesty and the Public. Edited by R. J. Lane, A.E.R.A. *Dedicated by Permission to Her Majesty the Queen.* 2 vols. Crown 8vo. 7s. 6d.
This is a careful reprint from the copy of Shakespeare used by Mr. Kemble at his public readings. The text is freed from all improprieties, and by judicious excisions each play is brought within the compass of an evening's reading.

DANTE'S DIVINE COMEDY. Translated by the Rev. Henry Francis Cary. With all the Author's Copyright Emendations. Post 8vo. 7s. 6d.

—— The Popular Edition, neatly Printed in Double columns. Royal 8vo. Sewed, 1s. 6d.; cloth, 2s. 6d.

PETRARCH'S SONNETS, TRIUMPHS, AND OTHER POEMS. Translated into English Verse. With Thomas Campbell's Life of the Poet. Illustrated. Post 8vo. 5s.

GOLDSMITH'S POEMS. Illustrated. 16mo. 2s. 6d.

WORDSWORTH'S WHITE DOE OF RYLSTONE; or, the Fate of the Nortons. Illustrated. 16mo. 3s. 6d.

LONGFELLOW'S EVANGELINE. Illustrated. 16mo. 3s. 6d.

—— WAYSIDE INN. Illustrated. 16mo. 3s. 6d.

Aldine Series of the British Poets.

Printed on toned paper in fcap. 8vo. size, and neatly bound in cloth gilt.

⁎ A cheap reprint of the series, so far as published, may now be obtained, cloth gilt, 1*s*. 6*d*. per volume.

AKENSIDE. With Memoir by the Rev. A. Dyce, and additional Letters. 5*s*.

BEATTIE. With Memoir by the Rev. A. Dyce. 5*s*.

BURNS. With Memoir by Sir Harris Nicolas, and additional Copyright Pieces. 3 vols. 15*s*.

BUTLER. With Memoir by the Rev. J. Mitford. 2 vols. 10*s*.

CHAUCER. Edited by R. Morris, with Memoir by Sir Harris Nicolas. 6 vols. £1 10*s*.

CHURCHILL. Took's edition, revised, with Memoir by James Hannay. 2 vols. 10*s*.

COLLINS. Edited, with Memoir by W. Moy Thomas. 3*s*. 6*d*.

COWPER, including his Translations. Edited, with Memoir, and Additional Copyright Pieces, by John Bruce, F.S.A. 3 vols. 15*s*.

DRYDEN. With Memoir by the Rev. R. Hooper, F.S.A. Carefully revised. 5 vols. £1 5*s*.

FALCONER. With Memoir by the Rev. J. Mitford. 5*s*.

GOLDSMITH. With Memoir by the Rev. J. Mitford. Revised. 5*s*.

GRAY. With Notes and Memoir by the Rev. John Mitford. 5*s*.

KIRKE WHITE. With Memoir by Sir H. Nicolas, and additional Notes. Carefully revised. 5*s*.

MILTON. With Memoir by the Rev. J. Mitford. 3 vols. 15*s*.

PARNELL. With Memoir by the Rev. J. Mitford. 5s.

POPE. With Memoir by the Rev. A. Dyce. 3 vols. 15s.

PRIOR. With Memoir by the Rev. J. Mitford. 2 vols. 10s.

SHAKESPEARE'S POEMS. With Memoir by the Rev. A. Dyce. 5s.

SPENSER. Edited, with Memoir, by J. Payne Collier. 5 vols. £1 5s.

SURREY. Edited, with Memoir, by James Yeowell. 5s.

SWIFT. With Memoir by the Rev. J. Mitford. 3 vols. 15s.

THOMSON. With Memoir by Sir H. Nicolas, annotated by Peter Cunningham, F.S.A., and additional Poems, carefully revised. 2 vols. 10s.

WYATT. Edited, with Memoir, by James Yeowell. 5s.

YOUNG. With Memoir by the Rev. J. Mitford, and additional Poems. 2 vols. 10s.

To be added to the Series shortly.

CHATTERTON. Edited by the Rev. W. W. Skeat, M.A., late Fellow of Christ College, Cambridge. Including the acknowledged Poems and Satires, the Rowley Poems, with an Essay proving their Authorship, a Memoir of the Poet, and Selections from his Prose Writings. In 2 vols.

CAMPBELL. Edited by his son-in-law, the Rev. A. W. Hill. With a Memoir by W. Allingham, Esq.

WILLIAM BLAKE. Edited, with a Memoir, by W. M. Rossetti.

ROGERS. With Memoir by Edward Bell.

THE COURTLY POETS, FROM RALEIGH TO MONTROSE. Containing the Complete Poetical Works of Sir Walter Raleigh, now first collected and authenticated, and the Poems of Sir Henry Wotton, with Selections from other Poets from 1540 to 1650. Edited by Rev. J. Hannah, D.C.L. With portrait of Raleigh.

Pocket Volumes.

A Series of Select Works of Favourite Authors, adapted for general reading, moderate in price, compact and elegant in form, and executed in a style fitting them to be permanently preserved. 32mo. Cloth, gilt tops.

Ready.

BACON'S ESSAYS. 2*s.* 6*d.*

BURNS'S POEMS, 3*s.*   } These editions contain all the
BURNS'S SONGS, 3*s.*   } copyright pieces published in
                    } the Aldine Edition.

COLERIDGE'S POEMS. 3*s.*

GEORGE HERBERT'S POEMS. 2*s.* 6*d.*

GEORGE HERBERT'S REMAINS. 2*s.*

GEORGE HERBERT'S WORKS. 3*s.* 6*d.*

LAMB'S TALES FROM SHAKESPEARE. 3*s.*

LONGFELLOW'S POEMS. 3*s.*

MILTON'S PARADISE LOST. 3*s.*

MILTON'S PARADISE REGAINED, and other Poems. 3*s.*

SEA SONGS AND BALLADS. By Charles Dibdin and others. 3*s.*

SOUTHEY'S LIFE OF NELSON. 3*s.*

THE LIEUTENANT AND COMMANDER. By Captain Basil Hall, R.N. 3*s.* 6*d.*

THE MIDSHIPMAN.—Sketches of his own early career, by the same Author. 3*s.* 6*d.*

THE ROBIN HOOD BALLADS. 3*s.*

WALTON'S LIVES OF DONNE, WOTTON, HOOKER, &c. 3*s.* 6*d.*

WALTON'S COMPLETE ANGLER. 3*s.*

WASHINGTON IRVING'S SKETCH-BOOK. 3*s.* 6*d.*

WASHINGTON IRVING'S TALES OF A TRAVELLER. 3*s.* 6*d.*

WHITE'S NATURAL HISTORY OF SELBORNE. 3*s.* 6*d.*

SHAKESPEARE'S PLAYS AND POEMS. Edited by T. Keightley. 13 vols. in handsome cloth case, 21*s.*

Elzevir Series of Standard Authors.

Small fcap. 8vo.

The publishers having been favoured with many requests that their Pocket Volumes should appear in a larger size, so as to be more suitable for Presents and School Prizes, are now printing New Editions in accordance with these suggestions.

They are issued under the general title of " Elzevir Series," to distinguish them from their other collections. This general title has been adopted to indicate the spirit in which they are prepared ; that is to say, with the greatest possible accuracy as regards text, and the highest degree of beauty that can be attained in the workmanship.

They are printed at the Chiswick Press, on fine paper, with rich margins, and are issued in tasteful bindings at prices varying from 3*s.* 6*d.* to 5*s.*

BURNS'S POEMS, with Portrait. 4*s.* 6*d.* } Containing all the copyright pieces published in the Aldine Edition.

BURNS'S SONGS. 4*s.* 6*d.*

COLERIDGE'S POEMS, with Portrait. 4*s.* 6*d.*

COWPER'S TABLE TALK AND MISCELLANEOUS POEMS. 4*s.* 6*d.*

COWPER'S TASK AND TRANSLATIONS. 4*s.* 6*d.*

IRVING'S SKETCH BOOK, with Portrait. 5*s.*

IRVING'S TALES OF A TRAVELLER. 5*s.*

LONGFELLOW'S POEMS, with Portrait. 4*s.* 6*d.*

LONGFELLOW'S SONG OF HIAWATHA AND GOLDEN LEGEND. 4*s.* 6*d.*

LONGFELLOW'S SPANISH STUDENT, and Later Poems. 4*s.* 6*d.*

MILTON'S PARADISE LOST, with Portrait. 4*s.* 6*d.*

MILTON'S PARADISE REGAINED, and other Poems. 4*s.* 6*d.*

SHAKSPEARE'S PLAYS AND POEMS. Edited by Thomas Keightley, 7 vols., with Portrait. 5*s.* each.

SOUTHEY'S LIFE OF NELSON, with Portrait. 4*s.* 6*d.*

WALTON'S ANGLER, with Engravings, 4*s.* 6*d.*

WALTON'S LIVES OF DONNE, HOOKER, &c., with Portrait. 5*s.*

c

Standard Works.

FIELDING (HENRY). Works, complete. With Memoir of the Author by Thomas Roscoe, and 20 Plates by Cruikshank. Medium 8vo. 14*s.*

—— The Novels separately. With Memoir by Thomas Roscoe, and Plates by Cruikshank. Medium 8vo. 7*s.* 6*d.*

SWIFT. Works of Jonathan Swift, D. D. Containing interesting and valuable passages not hitherto published, with Memoir of the Author by Thomas Roscoe. 2 vols. Medium 8vo. 24*s.*

SMOLLETT. Miscellaneous Works of Tobias Smollett. Complete in 1 vol. With Memoir of the Author by Thomas Roscoe. 21 Plates by Cruikshank. Medium 8vo. 14*s.*

LAMB. The Works of Charles Lamb, with a Memoir by Sir Thomas Noon Talfourd. Imp. 8vo. 10*s.* 6*d.*

HEYWOOD. The Proverbs of John Heywood. Edited, with Notes and Introduction, by Julian Sharman.

ASCHAM. The Schole Master. By Roger Ascham. Edited, with copious Notes and a Glossary, by the Rev. J. E. B. Mayor, M.A. 6*s.*

LOCKE ON THE CONDUCT OF THE HUMAN Understanding. Edited by Bolton Corney, Esq., M.R.S.L. 3*s.* 6*d.*

"I cannot think any parent or instructor justified in neglecting to put this little treatise into the hands of a boy about the time when the reasoning faculties become developed."—*Hallam.*

Recent Poetical Works.

ALLINGHAM. Day and Night Songs; and the Music Master. A Love Poem. By William Allingham. With nine Woodcuts, seven designed by Arthur Hughes, one by D. G. Rossetti, and one by John E. Millais, R.A. Fcap. 8vo. 5s.

—— Fifty Modern Poems. By W. Allingham, Author of "Day and Night Songs," and "Laurence Bloomfield." Fcap. 8vo. 5s.

—— Choice Lyrics and Poems ; or, Nightingale Valley. A Collection from the Time of Shakespeare to the Present Day. Edited by William Allingham. Fcap. 8vo. 3s. 6d.

BAILEY (P. J.) Festus: a Poem. By Philip James Bailey. *Eighth edition, revised.* Crown 8vo. 5s. With Portrait of Author, 6s.

—— Universal Hymn. By Philip James Bailey, Author of "Festus." Small 4to. Cloth. 5s.

BARRY CORNWALL. English Songs and Lyrics. *New edition.* Fcap. 8vo. 6s.

CALVERLEY (C. S.) Verses and Translations. *Fourth edition.* Fcap. 8vo. 5s.

—— Translations into English and Latin. Crown 8vo. 7s. 6d.

—— Theocritus Translated into English Verse. Crown 8vo. 7s. 6d.

—— Fly Leaves. *Third edition.* Fcap. 8vo. 3s. 6d.

CANONESS (THE). A Tale in Verse of the Time of the First French Revolution. Small 8vo. 5s.

CONINGTON. The Odes and Carmen Seculare of Horace translated in English Verse by the late John Conington, M.A. *Fifth edition.* Fcap. 8vo. 5s. 6d.

—— The Satires and Epistles of Horace translated into English Verse. *Third edition.* 6s. 6d.

CROLY (J. W.) Old Jewels Reset. Fables in Verse.
By J. W. Croly; with Twenty-five Illustrations by J. Proctor.
Demy 8vo. 10s. 6d.

FERGUSON (S.) Congal: a poem in Five Books by
Samuel Ferguson. Fcap. 4to. Handsomely bound. 12s.

GREENWELL (DORA). Carmina Crucis. By Dora
Greenwell. Post 8vo. With Six Illustrations by G. D. Leslie,
A.R.A. 5s.

> " In nobility and height of aim, Miss Greenwell's ' Carmina Crucis ' are
> honourably distinguished from the great mass of verse on either secular or
> sacred subjects, and it follows almost of necessity from the exalted dignity
> of their conception that the execution lags a little behind it. . . . Her
> poems have the noble fault of containing more breadth of intellectual view
> and depth of spiritual insight than she is capable of combining with
> artistic symmetry. Yet even this modified praise seems too much like
> censure when we read a striking passage like the following, extracted from
> her poem on ' Election :'—
>> ' Who shall the secret learn
>> Of thine exclusion stern ?' &c.
> After quoting these profound and moving lines, there is, perhaps, an anti-
> climax in observing that the vignettes of Miss Greenwell's volume have
> qualities of expression that will recommend them to really artistic minds ;
> to untrained eyes they may seem only inconspicuous and coarsely executed
> woodcuts. Such ornaments are well mated to these remarkable poems."—
> *Guardian.*

KEARY (E.) Little Sealskin and other Poems.
In the press.

MONSELL (DR.) Lights and Shadows, or Double
Acrostics. By the Old Vicar. *Third edition.* 1s. 6d.

—— Hymns of Love and Praise for the Church's Year.
By the Rev. J. S. B. Monsell, LL.D., Rector of St. Nicholas,
Guildford. *Second edition.* Fcap. 8vo. 3s. 6d.

—— The Passing Bell, and other Poems. By the Rev. J.S.
B. Monsell, LL.D., Rector of St. Nicholas, Guildford. *New
edition.* 32mo. 2s. 6d.

PRIDEAUX (MRS.) The Nine Days' Queen. A Dra-
matic Poem. By Mrs. Frederick Prideaux, Author of "Claudia."
Fcap. 8vo. 5s.

PROCTER (ADELAIDE ANNE). Legends and Lyrics.
Illustrated edition. With Portrait and Introduction. 21s.

—— Legends and Lyrics. First Series. *Fifteenth edition*, with Additional Poems. Introduction by Charles Dickens,
and Portrait of the Author. Fcap. 6s.

—— The Second Series. *Ninth edition*. Fcap. 8vo. 5s.

ROGERS (MISS). My Vis-à-Vis, and other Poems. By Mary Eliza Rogers, Author of "Domestic Life in Palestine." Small 8vo. 4s.

STORY OF QUEEN ISABEL (THE), and other Verses. By M. S., Author of "Twice Lost," and "The Linnet's Trial." 3s. 6d.

TREVELYAN (G. O.) The Ladies in Parliament, and other Pieces. Republished with Additions and Annotations. By G. O. Trevelyan, M.P., late Scholar of Trinity College, Cambridge. 6s. 6d.

> CONTENTS :—The Ladies in Parliament—Horace at Athens The Cambridge Dionysia—The Dawk Bungalow—A Holiday among some Old Friends.

WAUGH (EDWIN). Poems and Lancashire Songs. By Edwin Waugh, Author of "Come whoam to thi Childer and Me." *Third edition, enlarged.* Fcap. 8vo. 7s. 6d.

THE EPIGRAMMATISTS. Selections from the Epigrammatic Literature of Ancient, Mediæval, and Modern Times. With Notes, Observations, Illustrations, and an Introduction. By the Rev. Henry Philip Dodd, M.A., of Pembroke College, Oxford. 10s. 6d.

History.

YER (Dr.)　The History of the Kings of Rome.
By T. Dyer, LL. D.　With a Prefatory Dissertation
on the Sources and Evidences of Early Roman History.
Demy 8vo.　16s.

LONG (G.)　The Decline of the Roman Republic.　From
the Destruction of Carthage to the Consulship of Julius Cæsar.
By George Long, M. A.　4 vols.　8vo.　14s. per vol.　*Vol.* 5
(*conclusion*) *in the Press.*

PEARSON (C. H.)　A History of England during the
Early and Middle Ages. By Charles H. Pearson, M. A., Fellow
of Oriel College, Oxford. *New edition*, much enlarged, in two
volumes. Vol. I. 8vo. 16s. Vol. II., To the Death of Edward
I.　8vo.　14s.

—— Historical Maps of England during the first
Thirteen Centuries.　With Explanatory Essays and Indices.
By Charles H. Pearson, M.A.　Imperial folio.　*Second Edition.*
£1 11s. 6d.

—— The Barons' War.　Including the Battles of Lewes
and Evesham. By William Henry Blaauw, Esq., M.A.
Second edition, with additions and corrections by C. H. Pearson,
M. A.　Demy 8vo.　10s. 6d.

STRICKLAND (AGNES).　Lives of the Queens of
England, from the Norman Conquest to the Reign of Queen
Anne.　By Agnes Strickland.　Abridged by the Authoress for
the use of Schools and Families.　Post 8vo.　6s. 6d.

—— Lives of the last Four Princesses of the Royal
House of Stuart.　By Miss Agnes Strickland, forming an
appropriate Sequel to the "Lives of the Queens of England."
With a photograph of the Princess Mary, after a picture by
Honthorst.　Crown 8vo.　12s.

MAXWELL (W. H.)　History of the Irish Rebellion in
1798.　By W. H. Maxwell.　With Portraits and Etchings on
Steel by George Cruikshank.　7s. 6d.

FERGUSON, (M. C.)　The Story of the Irish before the
Conquest.　From the Mythical Period to the Invasion under
Strongbow.　By M. C. Ferguson.　Fcap.　5s.

Biography.

COOPER (THOMPSON). A New Biographical
Dictionary. By Thompson Cooper, F.S.A., Joint
Editor of "Athenæ Cantabrigienses." 1 vol. 8vo.
12s.

This volume is not a mere repetition of the contents of pre-
vious works, but embodies the results of many years' laborious
research in rare publications and unpublished documents. It is
intended to publish an Appendix, to be enlarged from time to
time so as to add the Biographies of distinguished persons who
have died during the progress of the work or since its publication. Any note of omission which may be sent to the Publishers
will be duly considered.

—— Athenæ Cantabrigienses. By C. H. Cooper,
F.S.A., and Thompson Cooper, F.S.A. Volume I. 1500—
1585. 8vo. Cloth. 18s. Volume II. 1586—1609. 18s.

BRYAN. A Biographical and Critical Dictionary of
Painters and Engravers. With a List of Ciphers, Monograms,
and Marks, by Michael Bryan. A *new edition, with numerous
additions*, by George Stanley. Imperial 8vo. £2 2s.

HELPS (SIR A.) The Life and Labours of the late Mr.
Brassey. By Sir Arthur Helps, K.C.B. *Dedicated by permission
to Her Majesty the Queen. New and cheaper edition*, 8vo. With
Illustrations. 10s. 6d. 5th thousand.

—— The Life of Hernando Cortes, and the Conquest of
Mexico. Dedicated to Thomas Carlyle. 2 vols. Crown 8vo.
15s.

—— The Life of Christopher Columbus, the Discoverer
of America. *Second edition.* Crown 8vo. 6s.

—— The Life of Pizarro. With Some Account of his
Associates in the Conquest of Peru. *Second edition.* Crown
8vo. 6s.

—— The Life of Las Casas, the Apostle of the Indies.
Second edition. Crown 8vo. 6s.

BONWICKE (AMBROSE), Life of, by his Father.
Edited by John E. B. Mayor, M.A., Fellow of St. John's
College, Cambridge. Small 8vo. 6s.

BEETHOVEN. Furioso, or Passages from the Life of Ludwig van Beethoven. From the German. Crown 8vo. 6s.

MACKENZIE (BISHOP), Memoir of the late. By the Bishop of Carlisle. With Maps, Illustrations, and an Engraved Portrait from a painting by G. Richmond. Dedicated by permission to the Lord Bishop of Oxford. *Second edition.* Small 8vo. 6s. The Large Paper Edition may still be had, price 10s. 6d.

PRYME (G.) Autobiographic Recollections of George Pryme, Esq., M.A., sometime Fellow of Trinity College, Professor of Political Economy in the University of Cambridge, and M.P. for the Borough. Edited by his Daughter, A. Bayne. 8vo. 12s.

BARBAULD (MRS.) A Memoir of Mrs. Barbauld, including letters and notices of her family and friends. By her great-niece, Anna Letitia Le Breton. With Portrait. Demy 12mo. 5s.

Law, Philosophy, &c.

ADAMS. The Law of Trade Marks. By F. M. Adams, Barrister-at-Law. 5s.

KENT'S COMMENTARY ON INTERNATIONAL LAW. *Revised*, with Notes and Cases brought down to the present year. Edited by J. T. Abdy, LL.D., Regius Professor of Laws in the University of Cambridge. 8vo. 16s.

TAYLOR (ISAAC). Elements of Thought. Post 8vo. 4s.

—— Home Education. Fcap. 8vo. 5s.

WHEWELL (DR.) Elements of Morality, including Polity. By W. Whewell, D.D., formerly Master of Trinity College, Cambridge. *Fourth edition.* In 1 vol. 8vo. 15s.

—— Lectures on the History of Moral Philosophy in England. By the Rev. W. Whewell, D.D. *New and Improved edition, with additional Lectures.* Crown 8vo. 8s.
The Additional Lectures are printed separately in Octavo for the convenience of those who have purchased the former Edition. 3s. 6d.

GROTE (J.) Exploratio Philosophica. Rough Notes on Modern Intellectual Science. Part I. -By the late J. Grote, B.D., Fellow of Trinity College and Professor of Moral Philosophy at Cambridge. 8vo. 9s.

—— An Examination of the Utilitarian Philosophy, by the late John Grote, B.D. Edited by Joseph B. Mayor, M.A., late Fellow of St. John's College, Cambridge. 8vo. 12s.

—— Aretaics and Eudæmonics, a Theory of Virtue and Happiness. By the late John Grote, B.D. *In the Press.*

GARVEY. A Manual of Human Culture. By M. A. Garvey, LL.B. Crown 8vo. 7s. 6d.

DRAPER (J. W) A History of the Intellectual Development of Europe. By John William Draper, M.D., LL.D. 2 vols. 8vo. £1 1s.

Social Advancement is as completely under the dominion of Natural Law as is bodily growth. The life of an individual is a miniature of the life of a nation. These propositions it is the special object of this book to demonstrate.

HELPS (Sir A.) Thoughts upon Government. By Sir Arthur Helps, K.C.B., Clerk of Her Majesty's Privy Council. Dedicated to the Earl of Derby. 8vo. 9s. 6d.

—— Brevia, or Short Essays and Aphorisms. By the author of "Friends in Council." *Second edition.* Crown 8vo. 6s.

—— Casimir Maremma. By the Author of "Friends in Council," "Realmah," &c. *Second and cheaper edition.* Post 8vo. 6s.

AN OLD MAN'S THOUGHTS ABOUT MANY THINGS. *Second and cheaper edition, with additions.* Small 8vo. 6s.

HARRIS (G.) Civilization considered as a Science, in Relation to its Essence, its Elements, and its End. *A New edition, with considerable Additions.* By George Harris, of the Middle Temple, Barrister-at-Law, F.S.A., author of "The Theory of the Arts," &c. Post 8vo. 6s.

CARNE (Miss). Country Towns, their Place in Civilization. By the Author of "Three Months' Rest at Pau." Fcap. 8vo. 3s.

BRASSEY (T.) On Work and Wages. By Thomas Brassey, M.P. *Third edition.* 8vo. 7s. 6d.

Also a cheap edition for distribution. 2s. 6d.

SICK AND IN PRISON. By the Author of Friends in Fur and Feathers. Fcap. 8vo. 2s. 6d.

SCOTT (Dr.) The Deaf and Dumb; their Education and Social Position. By Dr. Scott, for thirty years Head Master of the West of England Institution for the Deaf and Dumb, Author of "Exercises for the Deaf and Dumb," "The Education of Imbeciles," &c. 8vo. Cloth. *Second edition, revised and enlarged.* 7s. 6d.

GILBART. The Principles and Practice of Banking. By the late J. W. Gilbart. *New edition, revised and adapted to present day* (1871). 8vo. 16s.

𝕿𝖍𝖊𝖔𝖑𝖔𝖌𝖞.

On the Old and New Testament and the Early Church.

ESTAMENTA XII PATRIARCHARUM; ad fidem Codicis Cantabrigiensis edita : accedunt Lectiones Cod. Oxoniensis. The Testaments of the XII Patriarchs : an attempt to estimate their Historic and Dogmatic Worth. By R. Sinker, M. A. Librarian of Trinity College, Cambridge. Small 8vo. 7s. 6d.

PEROWNE (CANON.) The Book of Psalms, a New Translation, with Introductions and Notes, Critical and Explanatory. By the Rev. J. J. Stewart Perowne, B. D., Canon Residentiary of Llandaff, and Fellow of Trinity College, Cambridge. 8vo. Vol. I. *Third Edition,* 18s.; Vol. II. *Third Edition, in the Press.*

—— The Book of Psalms—abridged Edition for Schools and Private Students. Crown 8vo. 10s. 6d.

GALLOWAY (W. B.) Isaiah's Testimony for Jesus. With an Historical Appendix, and Copious Tabular View of the Chronology, from the Original Authorities. By W. B. Galloway, M. A., Incumbent of St. Mark's, Regent's Park, and Chaplain to the Right Hon. Viscount Hawarden. 8vo. 14s.

EWALD. The Prophet Isaiah. Chapters I.-XXXIII. From the German of H. Ewald. By Octavius Glover, B. D. Crown 8vo. 6s.

CRUDEN'S CONCORDANCE TO THE OLD AND NEW TESTAMENT, or an Alphabetical and Classified Index to the Holy Bible, specially adapted for Sunday School Teachers, containing nearly fifty-four thousand references. Thoroughly revised and condensed by G. H. Hannay. Fcap. 2s.

SCRIVENER (DR.) Novum Testamentum Græcum, Textus Stephanici, 1550. Accedunt variæ lectiones editionum Bezæ, Elzeviri, Lachmanni, Tischendorfii, et Tregellesii. Curante F. H. Scrivener, M.A., LL.D. 16mo. 4s. 6d. This Edition embodies all the readings of Tregelles, and of Tischendorf's Eighth or latest Editions. An Edition with wide Margin for Notes. 7s. 6d.

SCRIVENER (Dr.) Codex Bezæ Cantabrigiensis. Edited, with Prolegomena, Notes and Facsimiles. By F. H. Scrivener, M.A. 4to. 26s.

—— A Full Collation of the Codex Sinaiticus with the Received Text of the New Testament ; to which is prefixed a Critical Introduction. By F. H. Scrivener, M. A. *Second edition, revised.* Fcap. 8vo. 5s.

"Mr. Scrivener has now placed the results of Tischendorf's discovery within the reach of all in a charming little volume, which ought to form a companion to the Greek Testament in the library of every Biblical student."—*Reader.*

—— An Exact Transcript of the Codex Augiensis, Græco-Latina Manuscript in Uncial Letters of S. Paul's Epistles, preserved in the Library of Trinity College, Cambridge. To which is added a Full Collation of Fifty Manuscripts containing various portions of the Greek New Testament deposited in English Libraries : with a full Critical Introduction. By F. H. Scrivener, M. A. Royal 8vo. 26s.

The Critical Introduction is issued separately, price 5s.

—— A Plain Introduction to the Criticism of the New Testament. With forty Facsimiles from Ancient Manuscripts. For the Use of Biblical Students. By F. H. Scrivener, M. A., Trinity College, Cambridge. 8vo. *New edition in the Press.*

ALFORD (Dean). The Greek Testament : with a critically revised Text ; a Digest of Various Readings ; Marginal References to Verbal and Idiomatic Usage ; Prologemena ; and a Critical and Exegetical Commentary. For the use of Theological Students and Ministers. By the late Henry Alford, D. D., Dean of Canterbury.

Vol. I. *Sixth edition,* containing the Four Gospels. £1 8s.

Vol. II. *Sixth edition,* containing the Acts of the Apostles, the Epistles to the Romans and Corinthians. £1 4s.

Vol. III. *Fifth edition,* containing the Epistle to the Galatians, Ephesians, Philippians, Colossians, Thessalonians,—to Timotheus, Titus, and Philemon. 18s.

Vol. IV. Part I. *Fourth edition,* containing the Epistle to the Hebrews, and the Catholic Epistles of St. James and St. Peter. 18s.

Vol. IV. Part II. *Fourth edition,* containing the Epistles of St. John and St. Jude, and the Revelation. 14s.

Vol. IV. Complete. £1 12s.

ALFORD (Dean). The New Testament for English Readers. Containing the Authorized Version, with additional corrections of Readings and Renderings ; Marginal References ; and a Critical and Explanatory Commentary. By Henry Alford, D. D. In two volumes.

> Vol. I. Part I. Containing the First Three Gospels. *Third edition.* 12*s.*
>
> Vol. I. Part II. Containing St. John and the Acts. *Second edition.* 10*s.* 6*d.*
>
> Vol. II. Part I. Containing the Epistles of St. Paul. *Second edition.* 16*s.*
>
> Vol. II. Part II. Containing the Epistle to the Hebrews, the Catholic Epistles, and the Revelation. *Second edition.* 16*s.*

BARRETT (A. C.) Companion to the Greek Testament. For the use of Theological Students and the Upper Forms in Schools. By A. C. Barrett, M. A., Caius College ; Author of " A Treatise on Mechanics and Hydrostatics." *Third edition, enlarged and improved.* Fcap. 8vo. 5*s.*

This volume will be found useful for all classes of Students who require a clear epitome of Biblical knowledge. It gives in a condensed form a large amount of information on the Text, Language, Geography, and Archæology ; it discusses the alleged contradictions of the New Testament and the disputed quotations from the old, and contains introductions to the separate books. It may be used by all intelligent students of the sacred volume ; and has been found of great value to the students of Training Colleges in preparing for their examinations.

BASS'S COMPLETE GREEK AND ENGLISH LEXICON TO THE NEW TESTAMENT. 2*s.*

GRIESBACH. The New Testament in Greek. Griesbach's Text, with the various readings of Mill and Scholz at foot of page, and Parallel References in the margin ; also a Critical Introduction and Chronological Tables. Two fac-similes of Greek Manuscripts. 650 pages. 3*s.* 6*d.* ; or with Lexicon, 5*s.*

SCHOLEFIELD (Prof.) Hints for some Improvements in the Authorised Version of the New Testament. By J. Scholefield, M. A., formerly Regius Professor of Greek in the University of Cambridge. *Fourth edition.* Fcap. 8vo. 4*s.*

BLUNT. Two Introductory Lectures on the Study of the Early Fathers delivered at Cambridge. By J. J. Blunt, D.D. 8vo. 4*s.* 6*d.*

TERTULLIANI LIBER APOLOGETICUS. The Apology of Tertullian. With English Notes and a Preface, intended as an Introduction to the Study of Patristical and Ecclesiastical Latinity. By H. A. Woodham, LL.D. *Second edition.* 8vo. 8*s.* 6*d.*

INDEX CANONUM, containing the Canons called Apostolical, the Canons of the undisputed General Councils, and the Canons of the Provincial Councils of Ancyra, Neo-Cæsarea, Gangra, Antioch and Laodicea, in GREEK AND ENGLISH. Together with a Complete Digest of the whole code of Canon Law in the undivided Primitive Church, alphabetically arranged. By John Fulton, D. D., Rector of Christ Church, Mobile, U.S.A. Demy 8vo. Cloth. 12*s.*

WRATISLAW. Life, Legend, and Canonization of St. John Nepomucen, Patron Saint and Protector of the Order of the Jesuits. By A. H. Wratislaw, M. A., Translator of "The Adventures of Baron Wenceslas Wratislaw, of Mitrowitz," "Diary of an Embassy." Crown 8vo. 3*s.*

COLET (DEAN). A Treatise in Latin on the Sacraments of the Church, by John Colet, D. D., with Introduction and Notes by J. H. Lupton, M. A. 8vo. 4*s.* 6*d.*

—— Two Treatises in Latin on the Hierarchies of Dionysius. By John Colet, D. D., formerly Dean of St. Paul's. Now first published, with a Translation, Introduction, and Notes, by J. H. Lupton, M. A., late Fellow of St. John's College, Cambridge. Demy 8vo. 12*s.*

—— An Exposition in Latin of St. Paul's Epistle to the Romans. By John Colet, D. D., formerly Dean of St. Paul's. Now first published with a Translation, Introduction, and Notes, by J. H. Lupton, M.A., late Fellow of St. John's College, Cambridge. Demy 8vo. 10*s.* 6*d.*

—— A Treatise on St. Paul's Epistle to the Corinthians. By John Colet. *In the Press.*

CHALLIS (J.) A Translation of the Epistle of the Apostle Paul to the Romans, with an Introduction and Critical Notes. By J. Challis, M. A., Fellow of Trinity College, and Plumian Professor of Astronomy, Cambridge. 8vo. 4*s.* 6*d.*

MASKEW (T. R.) Annotations on the Acts of the Apostles. Original and Selected. For the use of Candidates for the Ordinary B. A. Degree, Holy Orders, &c. With Examination Papers. By T. R. Maskew, M. A. *Second edition, enlarged.* 12mo. 5*s.*

LEWIN (T.) A Life of St. Paul. By Thomas Lewin, M. A., Trinity College, Oxford, Barrister-at-Law, F. S. A. Second edition, revised and greatly enlarged. Illustrated with numerous fine Engravings on Wood, Maps, and Plans. 2 vols. demy 4to. *In the Press.*

—— Fasti Sacri; or, a Key to the Chronology of the New Testament. 4to. 21*s.*

—— Siege of Jerusalem by Titus. With the Journal of a recent Visit to the City and a General Sketch of the Topography of Jerusalem from the Earliest Times down to the Siege. Demy 8vo. 10*s. 6d.*

On the Prayer Book, Service and Liturgy of the English Church.

THE BOOK OF COMMON PRAYER. Ornamented with Head-pieces and Initial Letters specially designed for this edition. Printed in red and black at the Cambridge University Press. 24mo. Best morocco. 10*s. 6d.* Also in ornamental bindings, at various prices.

HUMPHRY (W. G.) An Historical and Explanatory Treatise on the Book of Common Prayer. By W. G. Humphry, B.D., late Fellow of Trinity College, Cambridge. *Third edition, revised and enlarged.* Fcap. 8vo. 4*s. 6d.*

—— The New Table of Lessons Explained, with the Table of Lessons and a Tabular Comparison of the Old and New Proper Lessons for Sundays and Holydays. By W. G. Humphry, B.D., Prebendary of St. Paul's, and Vicar of St. Martin's-in-the-Fields, Westminster. Fcap. 1*s. 6d.*

—— The Student's Book of Common Prayer. With an Historical and Explanatory Treatise. By William Gilson Humphry, B.D., Vicar of St. Martin's-in-the-Fields. 24mo. 7*s. 6d.*

JEWELL (Bp.) Apology for the Church of England, with his famous Epistle on the Council of Trent, and a Memoir. 32mo. 2*s.*

WELCHMAN (ARCHD.) The Thirty-nine Articles of the Church of England, Illustrated with Notes, and confirmed by Texts of the Holy Scripture, and Testimonies of the Primitive Fathers, together with References to the Passages in several Authors, which more largely explain the Doctrine contained in the said Articles. By the Ven. Archdeacon Welchman. *New edition.* Fcap. 8vo. Interleaved for Students, 3s.

HARDWICK (C. H.) History of the Articles of Religion. To which is added a Series of Documents from A.D. 1536 to A.D. 1615. Together with illustrations from contemporary sources. By the late Charles Hardwick, B.D., Archdeacon of Ely. *Second edition, corrected and enlarged.* 8vo. 12s.

LUMBY (J. R.) History of the Creeds. By J. Rawson Lumby, M.A. Tyrwhitt's Hebrew Scholar, Crosse Divinity Scholar, Classical Lecturer of Queens', and late Fellow of Magdalene College, Cambridge. Crown 8vo. 7s. 6d.

—— The Ancient Confessions of the Sixteenth Century, with special reference to the Articles of the Church of England. By J. Rawson Lumby. *[In the Press.*

POWELL (T. E.) The Holy Feast; or, the Witness of Scripture to the Teaching of the Church of England concerning the Sacrament of the Lord's Supper. By the Rev. T. E. Powell, M.A., Vicar of Bisham. Post 8vo. Cloth. 2s.

THE RECTOR AND HIS FRIENDS. Dialogues on some of the Leading Religious Questions of the Day. Crown 8vo. 7s. 6d.

MONSELL (DR.) Our New Vicar; or, Plain Words on Ritual and Parish Work. By the Rev. J. S. B. Monsell, LL.D., author of "Hymns of Love and Praise," &c. *Fifth edition.* Fcap. 8vo. 5s.

CARLISLE (BP. OF). A Guide to the Parish Church. By the Rt. Rev. Harvey Goodwin, D.D. 1s. sewed; 1s. 6d. cloth.

BROUGHTON (BP.) Sermons on the Church of England; its Constitution, Mission, and Trials. By the Rt. Rev. Bishop Broughton. Edited, with a Prefatory Memoir, by the Ven. Archdeacon Harrison. 8vo. 10s. 6d.

THE ENGLISH CHURCHMAN'S SIGNAL. By the Writer of "A Plain Word to the Wise in Heart." Fcap. 8vo. 2s. 6d.

A PLAIN WORD TO THE WISE IN HEART ON OUR DUTIES AT CHURCH, AND ON OUR PRAYER BOOK. *Fourth edition.* Sewed. 1s.

PEARSON (Bp.) ON THE CREED. Carefully printed from an Early Edition. With Analysis and Index. Edited by E. Walford, M.A. Post 8vo. 5s.

TEMPLE (HENRY). The Catholic Faith ; or, What the Church Believes, and Why. Being Six Lectures on the Athanasian Creed, preached in the Church of St. John the Evangelist, Leeds. 12mo. pp. 121. Cloth, 2s.

CARLISLE (Bp. of). Commentaries on the Gospels, intended for the English Reader, and adapted either for Domestic or Private Use. By the Rt. Rev. H. Goodwin, D.D. Crown 8vo. S. Matthew, 12s. S. Mark, 7s. 6d. S. Luke, 9s.

DENTON (W.) A Commentary on the Gospels for the Sundays and other Holy Days of the Christian Year. By the Rev. W. Denton, A.M., Worcester College, Oxford, and Incumbent of St. Bartholomew's, Cripplegate. *New edition.* 3 vols. 8vo. 54s.

> Vol. I. Advent to Easter. 18s.
>
> Vol. II. Easter to the Sixteenth Sunday after Trinity. 18s.
>
> Vol. III. Seventeenth Sunday after Trinity to Advent ; and Holy Days. 18s.

—— Commentary on the Epistles for the Sundays and other Holy Days of the Christian Year. By the Rev. W. Denton, Author of " A Commentary on the Gospels," &c.

> Vol. I. Advent to Trinity. 8vo. 18s.
>
> Vol. II. Completing the work. 18s.

—— A Commentary on the Acts of the Apostles.
[*In the Press.*

—— The Grace of the Ministry. Considered as a Divine Gift of uninterrupted Transmission and Two-fold Character. Edited by the Rev. William Denton, M.A. Oxon, Author of "Commentary on the Gospels and Epistles," &c. 8vo. 16s.

D

SADLER (M. F.) The Second Adam and the New Birth; or, the Doctrine of Baptism as contained in Holy Scripture. *Fourth edition, greatly enlarged.* Fcap. 8vo. 4s. 6d.

> "The most striking peculiarity of this useful little work is that its author argues almost exclusively from the Bible. We commend it most earnestly to clergy and laity, as containing in a small compass, and at a trifling cost, a body of sound and Scriptural doctrine respecting the New Birth which cannot be too widely circulated."—*Guardian.*

—— The Sacrament of Responsibility; or, Testimony of the Scripture to the teaching of the Church on Holy Baptism, with especial reference to the Cases of Infants; and Answers to Objections. *Sixth edition.* 6d.

—— The Sacrament of Responsibility. With the addition of an Introduction, in which the religious speculations of the last twenty years are considered in their bearings on the Church doctrine of Holy Baptism, and an Appendix giving the testimony of writers of all ages and schools of thought in the Church. On fine paper, and neatly bound in cloth. 2s. 6d.

—— Church Doctrine—Bible Truth. *Fifth Edition.* 10th thousand. Fcap. 8vo. 3s. 6d.

This work contains a full discussion of the so-called Damnatory Clauses of the Athanasian Creed. The new edition has additional Notes on Transubstantiation and Apostolical Succession.

> "Some writers have the gift of speaking the right word at the right time, and the Rev. M. F. Sadler is pre-eminently one of them. 'Church Doctrine—Bible Truth,' is full of wholesome truths fit for these times . . . He has the power of putting his meaning in a forcible and intelligible way, which will, we trust, enable his valuable work to effect that which it is well calculated to effect, viz., to meet with an appropriate and crushing reply one of the most dangerous misbeliefs of the time."—*Guardian.*
>
> "A Manual of Church Doctrine, well-nigh complete in all its parts, evolved from Holy Scripture in that convincing method which Mr. Sadler may be said in his previous publications to have made his own."—*Ecclesiastic.*

—— A Series of Church Tracts. *In preparation.*

—— The Communicant's Manual; being a Book of Self-Examination, Prayer, Praise and Thanksgiving. By the Rev. M. F. Sadler. Royal 32mo. Cloth, 1s. 6d.

> **** A Cheap Edition in limp cloth, 8d.

——— A Larger Edition on fine paper, red rubrics. Fcap. 8vo. *Nearly ready.*

For Confirmation Classes.

SADLER (M. F.) The Church Teacher's Manual of Christian Instruction. Being the Church Catechism expanded and explained in Question and Answer, for the use of Clergymen, Parents and Teachers, by the Rev. M. F. Sadler, Author of "Church Doctrine—Bible Truth," "The Sacrament of Responsibility," &c. *Third edition, revised.* Fcap. 8vo. 2s. 6d.

"It is impossible to overrate the service to religious instruction achieved by this compact and yet pregnant volume. . . . We owe many boons to Mr. Sadler, whose sermons and theological lectures and treatises have wrought much good in matters of faith. This Catechetical Manual is second to none of such."—*English Churchman.*

YOUNG (P.) Lessons on Confirmation. By the Rev. Peter Young, M.A., Author of "Daily Readings for a Year on the Life of our Lord Jesus Christ." Fcap. 8vo. 2s. 6d.

BARRY (DR.) Notes on the Catechism. For the Use of Schools. By the Rev. Alfred Barry, D.D., Principal of King's College, London. *Second edition, revised.* Fcap. 2s.

BOYCE (E. J.) Catechetical Hints and Helps. A Manual for Parents and Teachers on giving instruction to Young Children in the Catechism of the Church of England. By Rev. E. J. Boyce, M.A. *Second edition.* Fcap. 2s.

MONSELL (DR.) The Winton Church Catechist. Questions and Answers on the Teaching of the Church Catechism. By Rev. J. S. B. Monsell, LL.D., Author of "Our New Vicar." Cloth, 3s. ; or in Four Parts, sewed, 9d. each.

Extract from a Letter of the Bishop of Winchester.
"Without pledging myself to every sentiment or interpretation of Scripture in it, I esteem it thoroughly Anglican and orthodox, carefully and lucidly written, and I trust it may be much blessed. Public catechising is now of more value than ever."—*Jan.* 28, 1874.

MILL (DR.) Lectures on the Catechism. Delivered in the Parish Church of Brasted, in the Diocese of Canterbury. By W. H. Mill, D.D. Edited by the Rev. B. Webb, M.A. Fcap. 8vo. 6s. 6d.

CARLISLE (BP. OF). Lectures upon the Church Catechism. By H. Goodwin, D.D., Bishop of Carlisle. 12mo. 4s.

—— Confirmation Day. Being a Book of Instruction for Young Persons how they ought to spend that solemn day, on which they renew the Vows of their Baptism, and are confirmed by the Bishop with prayer and the laying on of hands. By H. Goodwin, D.D., Bishop of Carlisle. Eighth Thousand. 2d., or 25 for 3s. 6d.

CARLISLE (Bp. of). Plain Thoughts concerning the Meaning of Holy Baptism. By H. Goodwin, D.D., Bishop of Carlisle. *Second edition.* 2*d.*, or 25 for 3*s.* 6*d.*

——— The Worthy Communicant ; or, "Who may come to the Supper of the Lord ?" By H. Goodwin, D.D., Bishop of Carlisle. *Second edition.* 2*d.*, or 25 for 3*s.* 6*d.*

BLUNT (J. S.) Life after Confirmation. 18mo. 1*s.*

BOYCE. Examination Papers on Religious Instruction. By Rev. E. J. Boyce, M.A. Sewed. 1*s.* 6*d.*

A SHORT EXPLANATION OF THE EPISTLES AND GOSPELS of the Christian Year, with Questions for Schools. Royal 32mo. 2*s.* 6*d.* ; calf, 4*s.* 6*d.*

KEMPTHORNE (J.) Brief Words on School Life. A Selection from Short Addresses based on a Course of Scripture reading in School. By the Rev. J. Kempthorne, late Fellow of Trinity College, Cambridge, and Head Master of Blackheath Proprietary School. Fcap. 3*s.* 6*d.*

CLARKE (J. E.) Children at Church. Twelve Simple Sermons. By Rev. J. Erskine Clarke. *New edition.* 2 vols. 1*s.* each ; together, in 1 vol. cloth gilt, 2*s.* 6*d.*

PAPERS ON PREACHING AND PUBLIC SPEAK-ING. By a Wykehamist. Second thousand. Fcap. 8vo. 5*s.*

GATTY (Dr.) The Bell ; its Origin, History, and Uses. By Rev. A. Gatty. 3*s.*

ELLACOMBE (H. T.) Practical Remarks on Belfries and Ringers. By the Rev. H. T. Ellacombe, M.A., F.A.S., Rector of Clyst St. George, Devonshire. *Second Edition*, with an Appendix on Chiming. Illustrated. 8vo. 3*s.*

WIGRAM (W.) Change Ringing Disentangled. By the Rev. Woolmore Wigram, M.A. 2*s.*

BARON (J.) Scudamore Organs, or Practical Hints respecting Organs for Village Churches and small Chancels, on improved principles. By the Rev. John Baron, M.A., Rector of Upton Scudamore, Wilts. With Designs by G. F. Street, F.S.A. *Second edition, revised and enlarged.* 8vo. 6*s.*

Devotional Works.

YOUNG (REV. P.) Daily Readings for a Year, on the Life of Our Lord and Saviour Jesus Christ. By the Rev. Peter Young, M.A. *Third edition, revised.* 2 vols. 8vo. £1 1s.

BLUNT (J. S.) Readings on the Morning and Evening Prayer and the Litany. By J. S. Blunt. *Third edition.* Fcap. 8vo. 3s. 6d.

HAWKINS (CANON). Family Prayers:—Containing Psalms, Lessons, and Prayers, for every Morning and Evening in the Week. By the late Rev. Ernest Hawkins, B.D., Prebendary of St. Paul's. *Eleventh edition.* Fcap. 8vo. 1s.

THE PARISH PRIEST'S BOOK OF OFFICES AND INSTRUCTIONS FOR THE SICK. Compiled by a Priest of the Diocese of Sarum. Post 8vo. 3s. 6d.

BROWNING (H. B.) Aids to Pastoral Visitation, selected and arranged by the Rev. H. B. Browning, M.A., Curate of St. George, Stamford. *Second edition.* Fcap. 8vo. 3s. 6d.

COMPTON (B.) Private Devotions for Church Helpers. By the Rev. B. Compton, Rector of All Saints', Margaret Street. 16mo. Cloth. 1s. 6d.

THOMAS À KEMPIS. On the Imitation of Christ. A New Translation. By H. Goodwin, D.D. *Third edition.* With fine Steel Engraving, after Guido, 5s.; without the Engraving, 3s. 6d. Cheap edition, 1s. cloth; 6d. sewed.

WHITMARSH (DR.) Forms of Sin; a Manual for Self-Examination. By the Rev. E. D. Whitmarsh, D.C.L. With a Frontispiece. Fcap. 4s. 6d.

CARTER (T. T.) The Devout Christian's Help to Meditation on the Life of Our Lord Jesus Christ. Containing Meditations and Prayers for every day in the year. Edited by the Rev. T. T. Carter, Rector of Clewer. 2 vols. Fcap. 8vo. 12s. Or in five parts, three at 2s. 6d.; and two at 2s. each.

MORRIS (J.) Book of Consolation in Sickness, Sorrow, Adversity and Old Age. Gathered from the Writings of the Wise and Good. By John Morris. Crown 8vo. Cloth, gilt edges. 6s.

TAYLOR (JEREMY). Holy Dying. Fcap. 8vo. 2s. 6d.

THE DEVOTIONAL LIBRARY.

Edited by the Very Rev. W. F. Hook, D. D.,
Dean of Chichester.

A Series of Works, original or selected from well-known Church
of England Divines, published at the lowest price, and suitable,
from their practical character and cheapness, for Parochial
distribution.

Short Meditations for Every Day in the Year. 2 vols.
(1,260 pages), 32mo. Cloth, 5s.; calf, gilt edges, 9s.

In Separate Parts.
Easter, 9d.; Trinity, Part I. 1s.; Trinity, Part II. 1s.

The Christian taught by the Church's Services. (490
pages), royal 32mo. Cloth, 2s. 6d.; calf, gilt edges, 4s. 6d.

In Separate Parts.
Advent to Trinity, cloth, 1s.; Trinity, cloth, 8d.; Minor
Festivals, 8d.

The History of Our Lord and Saviour Jesus Christ. In
Three Parts. With suitable Meditations and Prayers. By W.
Reading, M. A. 32mo. Cloth, 2s.

Devotions for Domestic Use. 32mo. Cloth, 2s.; Containing :—

> The Common Prayer Book, the best Companion in the Family
> as well as in the Temple. 3d.

> Litanies for Domestic Use. 2d.

> Family Prayers; or, Morning and Evening Services for every
> Day in the Week. By the late Bishop Hamilton. Cloth,
> 6d.; calf, 2s.

> Bishop Hall's Sacred Aphorisms. Selected and arranged
> with the Texts to which they refer. By the Rev. R. B.
> Exton, M.A. Cloth, 9d.; calf, 2s. 3d.

> *⁎* These are arranged together as being suitable for Domestic
> Use; but they may be had separately at the prices affixed.

HOOK'S (DEAN) DEVOTIONAL LIBRARY (*continued*).

Helps to Daily Devotion. 32mo. Cloth. 8*d*. Containing :—

> The Sum of Christianity. 1*d*.
> Directions for Spending One Day Well. ½*d*.
> Helps to Self-Examination. ½*d*.
> Short Reflections for Morning and Evening. 2*d*.
> Prayers for a Week. 2*d*.

Aids to a Holy Life. First Series. 32mo. Cloth, 1*s*. 6*d*.
Containing :—

> Prayers for the Young. By Dr. Hook. ½*d*.
> Pastoral Address to a Young Communicant. By Dr. Hook. ½*d*.
> Helps to Self-Examination. By W. F. Hook, D.D. ½*d*.
> Directions for Spending One Day Well. By Archbishop Synge. ½*d*.
> Rules for the Conduct of Human Life. By Archbishop Synge. 1*d*.
> The Sum of Christianity, wherein a short and plain Account is given of the Christian Faith ; Christian's Duty ; Christian Prayer ; Christian Sacrament. By C. Ellis. 1*d*.
> Ejaculatory Prayer ; or, the Duty of Offering up Short Prayers to God on all Occasions. By R. Cook. 2*d*.
> Prayers for a Week. From J. Sorocold. 2*d*.
> Companion to the Altar ; being Prayers, Thanksgivings, and Meditations. Edited by Dr. Hook. Cloth, 6*d*.
> *** Any of the above may be had for distribution at the prices affixed ; they are arranged together as being suitable for Young Persons and for Private Devotion.

Aids to a Holy Life. Second Series. 32mo. Cloth, 2*s*.
Containing :—

> Holy Thoughts and Prayers, arranged for use on each day of the week. 3*d*.
> The Retired Christian exercised on Divine Thoughts and Heavenly Meditations. By Bishop Ken. 3*d*.
> Penitential Reflections for Lent and the Days of Fasting, &c. 6*d*.
> The Crucified Jesus, a Devotional Commentary on Luke 22 and 23. By A. Horneck, D.D. 3*d*.
> Short Reflections for every Morning and Evening in the Week. By N. Spinckes. 2*d*.
> The Sick Man Visited ; or, Meditations and Prayers for the Sick Room. By N. Spinckes. 3*d*.
> *** These are arranged together as being suitable for private meditation and prayer ; they may be had separately at the prices affixed.

HOOK'S (DEAN) DEVOTIONAL LIBRARY (*continued*).

Devout Musings on the Book of Psalms. 2 vols. 32mo. Cloth, 5*s.*; calf antique, 12*s.* Or, in Four Parts, cloth, 1*s.* each.

Other Editions.

Short Meditations for Every Day in the Year. *New edition, carefully revised.* 2 vols. Fcap. 8vo. Large type. 14*s.*

The Christian Taught by the Church's Services. 1 vol. Fcap. 8vo. Large type. 6*s.* 6*d.*

Holy Thoughts and Prayers, arranged for Daily Use on each Day of the Week, according to the stated Hours of Prayer. *Fifth edition, with additions.* 16mo. Cloth, red edges, 2*s.*; calf, gilt edges, 3*s.*

A Companion to the Altar.. Being Prayers, Thanksgivings, and Meditations, and the Office of the Holy Communion. *Second edition.* Handsomely printed in red and black. 32mo. Cloth, red edges, 2*s.*; morocco, 3*s.* 6*d.*

The Church Sunday School Hymn Book. Edited by W. F. Hook, D.D., Dean of Chichester. *Large paper.* Cloth, 1*s.* 6*d.*; calf, gilt edges, 3*s.* 6*d.*
   Cheap edition. Cloth, 32mo. 8*d.*; calf, gilt. 2*s.* 6*d.*

Verses for Holy Seasons. By C. F. Alexander. Edited by the Very Rev. W. F. Hook, D.D. *Fifth edition.* Fcap. 3*s.* 6*d.*

PEARSON (C. B.) Sequences from the Sarum Missal ; with English Translations. By C. B. Pearson, Prebendary of Sarum and Rector of Knebworth. Fcap. 8vo. Cloth gilt, red edges, 6*s.*

MONSELL (DR.) Hymns of Love and Praise for the Church's Year. By the Rev. J. S. B. Monsell, LL.D. *Second edition.* Fcap. 8vo. 3*s.* 6*d.*

—— The Parish Hymnal ; after the Order of the Book of Common Prayer. By the Rev. J. S. B. Monsell, Rector of St. Nicholas, Guildford. Cloth, 32mo. 1*s.* 4*d.*

THE CHURCH HYMNAL (with or without Psalms). 12mo. Large type, 1*s.* 6*d.* 18mo., 1*s.* 32mo. for Parochial Schools, 6*d.*

HOLY SONGS FOR ALL SEASONS. Demy 16mo. Cloth, red edges, 2*s.*; limp cloth, cut edges, 1*s.* 6*d.*; paper wrapper, 1*s.*

Sermons.

BADGER (G. P.) The State of the Dead. Sermons by the Rev. George Percy Badger, late Chaplain in the Diocese of Bombay. *Second edition.* Fcap. 8vo. 3s. 6d.

This work is new to the English public, as the First Edition was published in Bombay.

BARRY (Dr.) Sermons Preached in the Chapel of Cheltenham College. By the Rev. Alfred Barry, D.D. Crown 8vo. 8s. 6d.

BENSON (Dr.) Work, Friendship, Worship. Three Sermons preached before the University of Cambridge in October and November, 1871. By E. W. Benson, D.D., Master of Wellington College ; late Fellow of Trinity-College, Cambridge ; Examining Chaplain to the Bishop of Lincoln. Small 8vo. 2s. 6d.

BONNEY (T. G.) Death and Life in Nations and Men. Four Sermons preached before the University of Cambridge in April, 1868. By T. G. Bonney, B.D., Fellow of St. John's College. 8vo. 3s. 6d.

BLENCOWE. Plain Sermons by the Rev. E. Blencowe. *Sixth edition.* Fcap. 8vo. 6s.

BLOMFIELD (Bp.) Twenty-four Sermons on Christian Doctrine and Practice. By C. J. Blomfield, D.D., late Lord Bishop of London. 8vo. 10s. 6d.

BLUNT (Dr.) Five Sermons Preached before the University of Cambridge in 1845 and 1847. By J. J. Blunt, D.D. 8vo. 5s. 6d.

BUTLER (Bp.) Sermons and Remains. With Memoir by the Rev. E. Steere, LL.D. 6s.

** This volume contains some additional remains, which are copyright, and render it the most complete edition extant.

—— Three Sermons on Human Nature, and Dissertation on Virtue. Edited by W. Whewell, D.D. With a Preface and a Syllabus of the Work. *Fourth and cheaper Edition.* Fcap. 8vo. 2s. 6d.

CARLISLE (BP. OF). Parish Sermons. By H. Goodwin, D.D. 1st Series. *Third edition.* 12mo. 6s.

—— Second Series. *Out of print.*

—— Third Series. Third edition. 12mo. 7s.

—— Fourth Series. 12mo. 7s.

—— Fifth Series. With Preface on Sermons and Sermon Writing. 7s.

—— Four Sermons Preached before the University of Cambridge in February, 1869. I. Parties in the Church of England. II. Use and Abuse of Liberty in the Church of England. III. The Message of the Spirit to the Church of England. IV. Discussions concerning the Holy Communion in the Church of England. By H. Goodwin, D.D., Bishop of Carlisle. Small 8vo. 4s.

—— Four Sermons Preached before the University of Cambridge in the Season of Advent, 1858. By H. Goodwin, D.D., Bishop of Carlisle. 12mo. 3s. 6d.

—— Christ in the Wilderness. Four Sermons Preached before the University of Cambridge in the month of February, 1855. By H. Goodwin, D.D., Bishop of Carlisle. 12mo. 4s.

—— The Ministry of Christ in the Church of England. Four Sermons Preached before the University of Cambridge. I. The Minister called. II. The Minister as Prophet. III. The Minister as Priest. IV. The Minister Tried and Comforted. By H. Goodwin, D.D., Bishop of Carlisle. Fcap. 8vo. 2s. 6d.

—— Doctrines and Difficulties of the Christian Religion contemplated from the Standing-point afforded by the Catholic Doctrine of the Being of our Lord Jesus Christ. Being the Hulsean Lectures for the year 1855. By H. Goodwin, D.D., Bishop of Carlisle. 8vo. 9s.

—— "The Glory of the Only Begotten of the Father seen in the Manhood of Christ." Being the Hulsean Lectures for the year 1856. By H. Goodwin, D.D., Bishop of Carlisle. 8vo. 7s. 6d.

DAVIES (J. L.) Sermons on Life in Christ. By the Rev. J. Llewellyn Davies, M.A., Rector of Christ Church, Marylebone. Fcap. 8vo. 5*s.*

FOWLE (T. W.) Types of Christ in Nature. Nine Sermons preached in the Parish Church of Staines. By the Rev. T. W. Fowle, M.A., Oxon, late Curate of the Parish, and now Curate in Charge of the Parish of Holy Trinity, Hoxton. Fcap. 8vo. 2*s.* 6*d.*

FRY (E.) The Doctrine of Election. An Essay. By Edward Fry. Crown 8vo. 4*s.* 6*d.*

GATTY (Dr.) Sermons. By the Rev. A. Gatty, D.D. 12mo. 8*s.*

—— Twenty Plain Sermons for Country Congregations and Family Reading. By the Rev. A. Gatty, D.D., Vicar of Eccles. Fcap. 5*s.*

GROTE (J.) Sermons by the late John Grote, B.D. Professor of Moral Philosophy in the University of Cambridge. Fcap. 8vo. 5*s.*

HOOK (Dean). Sermons Suggested by the Miracles of our Lord and Saviour Jesus Christ. By the Very Rev. Dean Hook. 2 vols. Fcap. 8vo. 12*s.*

MILL (Dr.) Sermons Preached in Lent, 1845, and on several former occasions, before the University of Cambridge. By W. H. Mill, D.D. 8vo. 12*s.*

—— Five Sermons on the Temptation of Christ our Lord in the Wilderness. Preached before the University of Cambridge in Lent, 1844. *New edition.* 8vo. 6*s.*

POTT (Archdeacon). Village Sermons. By the Rev. Alfred Pott, B.D., Archdeacon of Berkshire. Fcap. 8vo. 3*s.* 6*d.*

PRITCHARD (C.) Analogies in the Progress of Nature and Grace. Four Sermons preached before the University of Cambridge, being the Hulsean Lectures for 1867 ; to which are added Two Sermons preached before the British Association. By C. Pritchard, M.A., President of the Royal Astronomical Society, late Fellow of St. John's College. 8vo. 7*s.* 6*d.*

PEROWNE (CANON).   Immortality.   Four Sermons
preached before the University of Cambridge. Being the Hulsean Lectures for 1868. By J. J. S. Perowne, B.D., Canon
of Llandaff, Vice-Principal and Professor of Hebrew in St.
David's College, Lampeter. 8vo.   7s. 6d.

SADLER (M. F.)   Parish Sermons.   Trinity to Advent.
Second edition.   6s.

—— Plain Speaking on Deep Truths.   Second edition.
6s.

—— Abundant Life, and other Sermons.   6s.

WINCHESTER (BP. OF).   Messiah as Foretold and
Expected.   A Course of Sermons relating to the Messiah, as
interpreted before the Coming of Christ.   Preached before the
University of Cambridge in the months of February and March,
1862.   By the Right Reverend E. Harold Browne, D.D., Lord
Bishop of Winchester. 8vo. 4s.

Miscellaneous Theological Works.

BUTLER (BP.)   Analogy of Religion, with Analytical
Introduction and copious Index. By the Rev. Dr. Steer. 3s. 6d.

CARLISLE (BP. OF).   Essays on the Pentateuch.   By
the Rt. Rev. H. Goodwin, D.D. Fcap. 8vo.   5s.

DREW (G. S.)   Bishop Colenso's Examination of the
Pentateuch Examined. By the Rev. G. S. Drew, Author of
"Scripture Lands," "Reasons of Faith." Crown 8vo. 3s. 6d.

EWALD (H.)   Life of Jesus Christ.   By H. Ewald.
Edited by Octavius Glover, B.D., Emmanuel College, Cambridge. Crown 8vo. 9s.

GIRDLESTONE (W. H.)   An Enquiry concerning
Prayers for the Dead. 8vo. pp. 88. Cloth. 3s.

GLOVER (O.)   Doctrine of the Person of Christ, an
Historical Sketch.   By Octavius Glover, B.D., Fellow of
Emmanuel College, Cambridge. Crown 8vo. 3s.

—— A Short Treatise on Sin, based on the Work of
Julius Müller. By O. Glover, B.D., Fellow of Emmanuel
College, Cambridge. Crown 8vo. 3s. 6d.

KING (DR.)   Thoughts and Suggestions on the Teaching of Christ. By the late William King, M.D., Cantab. Crown
8vo. 6s. 6d.

MILL (DR.) Observations on the attempted Application of Pantheistic Principles to the Theory and Historic Criticism of the Gospels. By W. H. Mill, D.D., formerly Regius Professor of Hebrew in the University of Cambridge. *Second Edition, with the Author's latest notes and additions.* Edited by his Son-in-law, the Rev. B. Webb, M.A. 8vo. 14*s*.

PATRICK (BP.) The Appearing of Jesus Christ. A short Treatise by Symon Patrick, D.D., formerly Lord Bishop of Ely, now published for the first time from the Original MS. Edited by II. Goodwin, D.D. 18mo. 3*s*.

RIGG (A.) The Harmony of the Bible with Experimental Physical Science. Four Lectures. By the Rev. Arthur Rigg, M.A., of Chester College. Fcap. 8vo. 2*s*. 6*d*.

SINKER (R.) The Characteristic Differences between the Books of the New Testament and the immediately preceding Jewish and the immediately succeeding Christian Literature, considered as an Evidence of the Divine Authority of the New Testament. By R. Sinker, M.A., Librarian of Trinity College, Cambridge. Small 8vo. 3*s*. 6*d*.

TAYLOR (ISAAC). Fanaticism. Post 8vo. 6*s*.

—— Loyola and Jesuitism. Post 8vo. 5*s*.

—— Natural History of Enthusiasm. 5*s*.

—— Physical Theory of Another Life. *New edition.* Post 8vo. 5*s*.

—— Saturday Evening. Post 8vo. 5*s*.

WILLIAMS (ROWLAND). Rational Godliness. After the Mind of Christ and the Written Voices of the Church. By the late Rowland Williams, D.D., Professor of Hebrew at Lampeter. Crown 8vo. 10*s*. 6*d*.

—— Paraméswara-jnyána-góshthí. A Dialogue of the Knowledge of the Supreme Lord, in which are compared the Claims of Christianity and Hinduism, and various Questions of Indian Religion and Literature fairly discussed. By the late Rowland Williams, D.D. 8vo. 12*s*.

WRATISLAW (A. H.) Notes and Dissertations, principally on Difficulties in the Scriptures of the New Covenant. By A. II. Wratislaw, M.A., Head Master of King Edward VI. Grammar School, Bury St. Edmunds, formerly Fellow and Tutor of Christ's College, Cambridge. 8vo. 7*s*. 6*d*.

Works of Reference and Miscellaneous Books.

STUDENT'S GUIDE (THE) to the University of Cambridge. *Third edition, revised and corrected in accordance with the recent regulations.* Fcap. 8vo.
[*In the Press.*

TABLES OF INTEREST, enlarged and improved; calculated at Five per cent. ; Showing at one view the Interest of any Sum, from £1 to £365 ; they are also carried on by hundreds to £1,000, and by thousands to £10,000, from one day to 365 days. To which are added, Tables of Interest, from one to twelve months, and from two to thirteen years. Also Tables for calculating Commission on Sales of Goods or Banking Accounts, from one-eighth to five per cent., with several useful additions, among which are Tables for calculating Interest on large sums for one day, at the several rates of four and five per cent. to £100,000,000. By Joseph King, of Liverpool. *Twenty-Fourth edition.* With a Table showing the number of days from any one day to any other day in the Year. 8vo. 7s. 6d.

DOUBLE ENTRY ELUCIDATED. By B. W. Foster. *Tenth edition.* 4to. 8s. 6d.
The design of this work is to elucidate the immutable principles of Double Entry, and to exemplify the art as it is actually practised by the most intelligent accountants at home and abroad.

A NEW MANUAL OF BOOK-KEEPING, combining the theory and practice, with specimens of a set of books. By Philip Crellin, Accountant. Crown 8vo. 3s. 6d.

A HANDY-BOOK OF RULES AND TABLES for Verifying Dates with the Christian Era, etc. Giving an account of the Chief Eras and Systems used by various Nations ; with easy Methods for determining the Corresponding Dates. By John J. Bond, Assistant Keeper of the Public Records. Crown 8vo. 7s. 6d.
The four years "before the common account called Anno Domini," noticed in the margins of many editions of the Gospels are now accounted for.
The difficulties in reconciling the historical dates and facts arising from the Roman system of reckoning with the Augustan era, introduced in the third century, which differed from the era of Augustus, used in the first and second centuries, are, it is believed, now removed by distinguishing the reckoning of one era from the other.

PARLIAMENTARY SHORT-HAND (Official System). By Thompson Cooper. Fcap. 8vo. 2s. 6d.

DAYS OF THE WEEK in Years Past, Present, and Future—a Perpetual Calendar. By John J. Bond, Assistant Keeper of the Public Records, &c. *2s. 6d.*

UNDER GOVERNMENT: an Official Key to the Civil Service, and Guide for Candidates seeking Appointments under the Crown. By J. C. Parkinson, Inland Revenue, Somerset House. *Fifth edition.* Crown 8vo. *2s. 6d.*

EXECUTORS' ACCOUNT-BOOK. *4s.*

HINTS TO MAID SERVANTS in Small Households, on Manners, Dress, and Duties. By Mrs. Motherly. Fcap. 8vo. *1s. 6d.*

THE HOUSEKEEPING BOOK, or Family Ledger. On an Improved Principle, by which an exact Account can be kept of Income and Expenditure; suitable for any Year, and may be begun at any time. With Hints on Household Management, Receipts, &c. By Mrs. Hamilton. 8vo. Cloth, *1s. 6d.*; sewed, *1s.*

SYNONYMS AND ANTONYMS of the English Language. Collected and Contrasted. By the late Ven. C. J. Smith, M.A. Post 8vo. *5s.*

SYNONYMS DISCRIMINATED. A Catalogue of Synonymous Words in the English Language, with their various Shades of Meaning, &c. Illustrated by Quotations from Standard Writers. By the Ven. C. J. Smith, M.A. Demy 8vo. *16s.*

Dictionaries.

BRYAN'S DICTIONARY OF PAINTERS. See p. 39.

COOPER'S BIOGRAPHICAL DICTIONARY. See p. 39.

DR. RICHARDSON'S NEW DICTIONARY OF THE ENGLISH LANGUAGE. Combining Explanation with Etymology, and copiously illustrated by Quotations from the Best Authorities. *New edition*, with a supplement containing additional Words and further Illustrations. In 2 vols. 4to. *£4 14s. 6d.* Half-bound in Russia, *£5 15s. 6d.* Russia, *£6 12s.*

The *Words*, with those of the same family, are traced to their origin. The *Explanations* are deduced from the primitive meaning through the various usages. The *Quotations* are arranged chronologically, from the earliest period to the present time.

The Supplement separately. 4to. *12s.*

An 8vo. edition, without the Quotations, *15s.* Half Russia, *20s.* Russia, *24s.*

Dr. Webster's Celebrated Dictionaries.

THE GUINEA DICTIONARY.

WEBSTER'S "NEW ILLUSTRATED" DICTION-
ARY of the English Language, in 1 vol. 4to. The Etymology
by Dr. C. F. Mahn, of Berlin. Containing nearly one thousand
six hundred pages, with Three Thousand Illustrations. Strongly
bound in cloth. 21*s.*

The peculiar features of this edition are, Fulness and Com-
pleteness, Scientific and Technical Words, Accuracy of Defini-
tion, Pronunciation, Etymology, Uniformity in Spelling, Quotations, The Synonyms, The Illustrations, Cheapness.

The volume contains 1,576 pages, and is sold for One Guinea.
It will be found, on comparison, to be one of the cheapest
volumes ever issued. It can also be had strongly half-bound in
calf, or half-bound in Russia.

"For the student of English etymologically, Wedgwood, Ed. Muller,
and Mahn's Webster are the best dictionaries. While to the general
student Mahn's Webster and Craig's 'Universal Dictionary' are most useful."—*Athenæum.*

"The best practical English Dictionary extant."—*Quarterly Review*
Oct. 1873.

THE HALF GUINEA DICTIONARY.

—— PEOPLE'S DICTIONARY of the English Lan-
guage, based on Webster's Large Dictionary, and containing all
English words now in use, with their pronunciation, derivation,
and meanings. In One Volume, large 8vo. containing more
than one thousand pages and Six Hundred Illustrations. 10*s. 6d.*

This edition contains, All Scientific Words important to non-
scientific readers, Important Phrases, with Explanations, Synonyms, Orthography, Pronunciation, A Glossary of Scottish
Words and Phrases, Vocabularies of Scriptural, Classical,
and Geographical Proper Names, a Vocabulary of Perfect and
allowable Rhymes, &c., &c.

—— COMPLETE DICTIONARY of the English Lan-
guage contains all that appears in the above Dictionaries, and
also a valuable Appendix and seventy pages of Illustrations
grouped and classified, rendering it a complete Literary and
Scientific Reference Book. 1 vol. 4to. Strongly bound in
cloth. £1 11*s. 6d.*

"The cheapest Dictionary ever published, as it is confessedly one of
the best. The introduction of small woodcut illustrations of technical and
scientific terms adds greatly to the utility of the Dictionary."—
Churchman.

CHISWICK PRESS:—PRINTED BY WHITTINGHAM AND WILKINS,
TOOKS COURT, CHANCERY LANE.